Idols to Incubators

Idols to Incubators
Reproduction theory through the ages

Julia Stonehouse

Scarlet Press

Published by Scarlet Press
5 Montague Road, London E8 2HN

97 96 95 94
5   4   3   2   1

British Library Cataloguing-in-Publication Data
A catalogue record for this book is available from
the British Library

ISBN   1 85727 052 5 pb
        1 85727 057 6 hb

Illustrations by Edwina Hannam

Designed and produced for Scarlet Press by
Chase Production Services, Chipping Norton, OX7 5QR

Typeset from author's disks by
Stanford DTP Services, Milton Keynes

Printed in Finland by WSOY

For Lily

'There's no chance for the welfare of the world unless the condition of women is improved. It is not possible for a bird to fly on one wing.'

Shrii Shrii A'nanda Mu'rti (P. R. Sarkar),
Supreme Expression, Part Two

Contents

Acknowledgements

As I sat down to write these acknowledgements I became acutely aware of how indebted I am to women, for different reasons. For making known the history of women, I am grateful to Marija Gimbutas, Rosalind Miles, Merlin Stone and Elizabeth Gould Davis in particular. The powerful role of women in pre-history has been made clear to me through the work of Asphodel P. Long, Ruth Green and Sue Lewis, amongst others. And the negative impact of incorrect reproduction theory in contemporary society has most clearly been shown by the excellent insight provided by Carol Delaney.

While researching the book I was introduced to Irene Coates who, on the other side of the world, has like me been stopped in her tracks by the significance of Karl Ernst von Baer's discovery of the mammalian ovum. For her friendship and support, I give thanks. I am also grateful to Vivienne Greenwell for transcribing the manuscript. And last, but by no means least, I thank my mother who has, over the many years it has taken to see this work in print, provided unfailing encouragement, love and support.

The author and publisher wish to thank the following for permission to reproduce from copyright material: *Autobiography of Dr Karl Ernst von Baer*, Science History Publications; Marija Gimbutas, *Language of the Goddess*, Thames & Hudson Ltd; Louis Ginzberg, *Legends of the Bible*, The Jewish Publication Society; Frances Dahlberg, *Woman the Gatherer*, Yale University Press; Carol Delaney, *Seed and the Soil: Gender and Cosmology in Turkish Village Society*, University of California Press; Margaret Ehrenberg, *Women in Prehistory*, British Museum Press; Paula Gunn Allen, *The Sacred Hoop*, Beacon Press; Bronislaw Malinowski, *The Sexual Life of Savages*, Routledge and Kegan Paul; Rosalind Miles, *The Women's History of the World*, Michael Joseph; Veronica Veen, *The Goddess of Malta: The Lady of the Waters and the Earth*, Inanna, 1992.

Introduction

Human beings have had reproduction theories ever since the first four-year-old looked up and asked 'Where do I come from?' Something has to be said. We tell our children they come from ovum and sperm, the female and male 'seeds' which fuse in mummy's tummy. Mother and father contribute equally to genetic inheritance, we say, and we even have the equation to prove it: 23+23=46. That's 23 DNA-carrying chromosomes from each parent making the total number of 46 with which a newly fertilised ovum starts life. The blueprint of life comes half from the mother, and half from the father, and both can take pleasure in the fact that their children carry genes which reflect them like a mirror into the future. In societies which are educated in modern reproduction theory, the infinite joy of parenthood can be shared equally by women and by men.

However, in many other cultures throughout the world today children are taught that parenthood is not shared, but belongs to the father alone. This is the old 'seed and soil' theory which dominated thinking for nearly five thousand years and maintained that human 'seed' – genetic inheritance – came from men who planted it in the 'soil', the womb, the woman, the incubator. Women nurtured what men created, and there was no such thing as 'mother'. As a theory, this was not scientifically disproved until the turn of the twentieth century, but in many parts of the world that are still unaware of modern reproduction theory, the male-seed idea continues to be the cornerstone of all social and religious organisation.

As a concept, parentage was made most poignant for me the day my child was born. I remember thinking, 'My DNA is in you. In you I endure outside myself. I am extended by you and you give me continuity. All that matters is that you live, breathe and feel. It doesn't matter how different from me you are or become. In a way, the more different, the greater will be my sense of

1

expansion. My joy will be further extended if you have children of your own.' I held my baby very close and thought 'She's mine,' with the deep satisfaction of knowing that our links can never be severed. Time and even death cannot erase them. The irrevocable deed is done. I am a parent. Nourishing the baby for nine months seemed a mere additional privilege to the larger joy of parenthood.

It is this, parenthood, which was denied to women for much of what we know, appropriately, as 'his-story'. Broadly speaking, the male-seed concept began around 3000 B.C. and continued more or less unchallenged until A.D. 1900. Because it is a concept that still defines and confines the lives of so many women living today, and because the rest of us live within its cultural shadow, this theory of *Man the seed, woman the incubator* is discussed first, in Part One of this book. But the male-seed concept is only one of three major reproductive theories which have influenced human behaviour, in addition to our own scientifically-based one.

Moving backwards in time, Part Two, *Woman the seed, man the stimulator* looks at an interface period, broadly speaking between 8000 and 3000 B.C., in which people believed the 'seed' of future generations lay somehow within the bodies of women and that men stimulated it to grow. This view was developed from the earliest reproduction theory that women reproduced on their own, without men, and this is discussed in Part Three, *The parthenogenetic woman*.

As we shall see, reproduction theory affects a child's relationships with its mother, father, life-partner, children, god and society. It helps explain much about what is happening in the world today, and much about what went on before. Indeed, I don't think we can understand human relationships and social organisation, past or present, until we put reproduction theory into the picture. Human beings are driven as much by ideas as they are by instinct, and ideas about reproduction drive human beings very hard indeed.

Although the ancient Greeks most emphatically became male-seedists, they did not start out that way and the concept had to be drummed into the doubters through the propaganda medium of tragic drama performed on civic occasions. In the *Eumenides*, first performed in 458 B.C., Apollo gets Orestes off the charge of matricide with this argument:

> The mother is not the true source of life.
> We call her mother, but she is more the nurse,
> The furrow where the seed is thrust.
> The thruster, the father, is the true parent:
> The woman but tends the growing plant
> With god's good grace, the host
> That shelters the stranger-guest.[1]

Matricide was proved an invalid charge because there was no such thing as a mother; she was just the 'nurse' or the 'furrow' who tended the 'stranger-guest'.

As this idea was confirmed by the Bible, and could not be disproved by science, thinking on reproduction had not changed much by the beginning of the nineteenth century A.D. In an encyclopedia of 1813, under 'Generation' we learn that 'till within a very few years ... every anatomist, and indeed every man who pretended to the smallest portion of medical science, was convinced that his children were no more related, in point of actual generation, to his own wife than they were to his neighbours'.[2]

It is not so difficult to come to this conclusion. All you have to do is make the connection between intercourse and the arrival of a baby nine months later and say to yourself 'What goes in must come out.' If you are going to think in terms of human 'seed', it makes sense to think in terms of a singular seed – like a grain of wheat or the pip of an olive – because we don't, after all, put two seeds next to each other in the ground and expect them to fuse and grow into a single plant. If we keep animals, we soon notice that no rams equal no lambs, and assume that whatever the male delivers into the womb must carry the essence of life. It is then only a question of degree. Aristotle said men deliver the 'impetus and form' (the energy and blueprint for the new life), and also the 'sentient soul' (the thinking, creative part), but seventeenth-century 'spermists' maintained that the sperm contained the physical body as well, in miniature unfolding form.

The ovary and whatever it contained was usually thought to be nutritive, food for the growing baby within. Galen and others referred to a female 'seed', but it was always impotent for some dubious reason or another. There were a few who challenged this view over the ensuing centuries, notably William Harvey, the discoverer of blood circulation who, in 1651, stated that *omne vivum ex ovo* – all animals are produced from eggs, including man. Like most people, Harvey knew that an egg becomes a live chick only if it has been fertilised by a cockerel, and he deduced from this that the male in some way confers life. But he also knew of 'wind-eggs' – unfertilised eggs which go on to produce live chickens – and stated that the female must therefore be credited with contributing 'form, species and soul'. However, because those wind-egg chickens were themselves infertile and unable to reproduce, Harvey was led to believe that the power of replication, the 'fructifying power', was conferred on the new life by the father.

Nevertheless, Harvey was a rare defender of the female role, saying that women might have a 'more powerful right to be called the agent' of reproduction than men. But he couldn't prove it and nobody was going to be able to prove it until Karl Ernst von Baer looked down his microscope and discovered the mammalian ovum in 1827. 'Taking a single look through the instrument', he wrote in his autobiography, 'I recoiled as if struck by lightning.'[3] As well he might. Reproduction theory would never be the same again.

The knee-jerk reaction of the male-seed theorists was to say the newly discovered ovum was just a more specialised form of nutriment, a sort of larder the sperm-seed falls into. But the evidence against them was soon to add up. During the 1820s and '30s scientists saw frog sperm fertilise eggs, and cellular division in newts, birds, rabbits and so on. Cell theory in 1838 helped define the lines along which they were thinking. By the 1850s, sperm had been observed entering the egg many times, but not yet into the mammalian ovum. In 1861 the ovum was proved unicellular in vertebrates; in 1875 the segmentation of mammalian ova was observed; in 1876 Hertwig saw that it was (sea urchin) sperm and ovum *nuclei* that fuse; and in 1879 Fol showed that it involved a *single* (rabbit) sperm. And so, by small steps, we reached an understanding. Meanwhile, the study of cells (cytology) was turning to chromosomal theory, with Weismann anticipating it in 1892 and Boveri proving it ten years later. In 1891 the accessory chromosomes were seen, and in 1902 they were shown to be the determinants of gender. Mendel's work was rediscovered and by 1909 we had the word 'gene', which Thomas Morgan Hunt used to describe the sub-units of chromosomes which proved, in 1915, to be responsible for identifiable, hereditary traits.

And there we had it, in theory at least. The shock of Darwin had worn off and although we were still only talking about fruit flies and garden peas, most of us were prepared to accept that we were talking about humans too. We knew our exact inheritance in 1956 when Tjio and Levan demonstrated that the usual number of chromosomes per person is 46, and we got photographic proof of human fertilisation in 1960. Yes, we were like sea urchins; Oskar Hertwig would have been pleased.

But poor Aristotle would have fallen off his pedestal if he'd known what we know now. To Aristotle it was 'plain that the female does not contribute semen to the generation of offspring', to us, for different reasons, it is plain that she does. The male 'seed' comprises a package of DNA which is located in the sperm-head which meets with an equal amount of DNA from the nucleus of the ovum, and reproduction begins. However, it is the ovum itself which reproduces. Only the ovum contributes RNA, which self-replicates, and mitochondria, self-replicating molecular factories which supply the energy and deal with the oxygen. Only the ovum contributes protoplasm, the complex semi-fluid substance which synthesises proteins; while it also seems to be the home of that strange power, one might even call it the 'life force', which pulls the chromosomes to opposite poles in the cell. No educated person today would call women's 'seed' impotent, let alone deny it exists.

Reproductively, women need men's DNA for 'genomic imprinting', as the geneticists call it. The male role is very specific, and limited to the nucleic DNA which, to the scientists working with it in the lab, is something very physical. William Harvey, who had tried to find semen in the uterus of does after coition and failed, had been forced to assume that men's 'subtle effluvia' was 'incorporeal' – that is, spiritual or invisible, depending on your point of view. It was about as vague as Aristotle's 'sentient soul' – the thinking, creative part. We now know that genes for intelligence are found on the X, female, chromosome, and that creation is not a male preserve, reproductively or otherwise. As for semen contributing the 'impetus and form', it provides only enough impetus for the sperm to reach the ovum and thereafter the impetus is provided by the mitochondria in the ovum; while the 'form' is provided – in terms of blueprint – by both male and female DNA equally, and in terms of physical body by the ovum's capacity to synthesise proteins. Indeed, the ovum is so reproductively capable it raises the question of whether the ovum needs the sperm at all.

Artificial parthenogenesis – the stimulation of ova by artificial means – proved rather successful when Jacques Loeb and Yves Delage started this field of enquiry at the turn of the twentieth century. The scientific community was shocked by the ease with which the eggs of sea urchins, frogs and even rabbits could be stimulated to produce live adults using 'weak acids, temperature shock, treatment with fat solvents (such as ether, alcohol and benzene), osmotic changes produced by sugar or urea solutions, ultraviolet light and even the educated prick of a needle'.[4] The offspring were all female and resembled their mothers. Although artificial parthenogenesis can be brought about using either chemical, physical or mechanical means, as far as I know it has not been much developed over the years. It is not difficult to get the ovum to develop to the point of cell division, implantation, and to the stage of beating heart, distinct bodily segments (somites) and forelimbs, but most conceptuses die at this point. The reason seems not so much to be because of a deficiency in the foetus, as in the functioning of the placenta.[5] For obvious ethical reasons, this work is not carried out on human ova although it is well known that left alone, unfertilised by sperm, some ova will spontaneously begin to develop. Given all this, it is not outside the realms of possibility that at some time in the distant future women will be able to reproduce themselves parthenogenetically, without men.

The first definite evidence we have that people understood men were involved in reproduction is a little stone relief carving from Çatalhöyük in Turkey, dated between 6000 and 5500 B.C. It shows a couple in embrace on the left and a woman holding a baby on the right. The message seems pretty clear: sex = babies. It is assumed that people knew this connection for some time before this sculpture was made, but for how long? Mostly the date is given as 8000 B.C. because before then, in terms of artefacts, the male is conspicuous by his absence. Whereas we have hundreds of images of women and women's sexual parts, there are hardly any images of men and only one (or possibly two) carved phalli. In the earliest days of human society, in ritual at least, men do not seem to have been very important. After 8000 B.C., the male begins to make an appearance, but in a secondary role to the female. Women still seem to be the revered gender, as we can see from the artefacts and even pieces of written evidence.

Between 8000 and 3000 B.C., then, men were known to be involved in reproduction, but there is no evidence that they were thought to be actual generators, deliverers of human 'seed'. On the contrary, the evidence points to a female-centred creativity and, I believe, a role for men as 'stimulators' of that creativity.

Intercourse does not have to be seen as the delivery of male seed, especially when people are used to thinking of reproduction as essentially feminine. During these early days people could have thought in terms of semen irrigating the seed within the body of a woman, stimulating it to germinate and grow. Or male reproductive stimulation could have been thought of in terms of semen delivering special food for the baby within, or as non-generative 'fertiliser'. The stimulatory effect could even be attributed to the physiological 'shock' caused by the action of intercourse itself. Whichever form it takes, in the *Woman the seed, man the stimulator* theory, woman is the creative partner, while man is merely the facilitator or helper.

In the period before 8000 B.C., it is most likely that people thought in terms of a *parthenogenetic woman*, the woman who reproduces without the help of a man. Perhaps they thought you plucked a baby from the body of a woman as you pluck a fruit from a tree or a root from the ground. Alternatively, they could have thought that reproduction is a joint venture between spirits and women, which is what the Trobriand Islanders of Melanesia think to this very day. According to them, men play no part in reproduction other than keeping the baby-passage dilated and lubricated, so the baby can come *out*. Men contribute no 'seed', no 'vital impulse', no 'active seminal power' and no 'genomic imprinting'. Reproduction is entirely a process between the spirits of dead ancestors and women. Mother is the only parent and, because paternity is not an issue, women's sexuality is very free. The Trobrianders, who are discussed in Chapter 15, illustrate that different interpretations can be made of the act of intercourse as it relates to reproduction. To them, sex and babies are not connected, and they have a whole logic system to back up their theory. Their logic is no more or less flawed than Aristotle's. It is different, that's all.

Indeed, incorrect thinking on reproduction has been a feature of human life from the earliest times to the beginning of the twentieth century. This isn't really surprising considering the difficulty in actually establishing the facts of life, and given the various deductions that can be drawn from the act of intercourse. However, once the connection between intercourse and babies had been made, it was perhaps only a matter of time before the early idea – that women create while men facilitate – would become the later idea: that men create while women facilitate. There was nothing to prove the point either way but, unfortunately for women, men's reproductive jealousy and 'logic' took over and claimed reproduction as essentially male. From being the earlier 'waterer' of female seed within, the man became the 'deliverer' of seed, and master of the world.

Women couldn't be denied a role in procreation, but theirs was to become the role of the incubator. In 1800, Erasmus Darwin, Charles' grandfather, was saying that men supplied the rudimental embryo while women provided the oxygen, food and nidus (or nest), and backed up his theory up with the logic that as men are physically larger than women they 'should contribute as much or more toward the reproduction of the species'. As women already made a contribution in terms of nourishment and warmth, it made sense (to him) that the other contribution – life itself – should come from men.

The legacy of this thinking is still very much with us today. Anthropologist Carol Delaney found it in contemporary Turkey:

> With seed, men appear to provide the creative spark of life, the essential identity of a child; while women, like soil, contribute the nurturant material that sustains it ... although they appear to go together naturally, they are categorically different, hierarchically ordered, and differentially valued ... The perceived creative, life-giving ability of men allies them symbolically with God, whereas the material sustenance provided by women associates them with what was created by God, namely the earth.[6]

Throughout history reproductive theory has led to ideas about the 'nature' of men and of women which, because they are designed by God, are difficult, indeed blasphemous, to challenge. Also, reproductive theory leads to logic and 'facts' which seem incontrovertible. If a child asks 'Where do I come from?', and you answer 'From father's seed', other ideas inevitably follow. If father carries the 'seed', he must have got it from the paternal grandfather, who got it from his father and so on, until one reaches the original source of that seed: the male God. If men are connected to the divine in this way, they are not only more creative than women, but more spiritual too. For the long line of creativity to be maintained, a man must produce sons. Girls are, reproductively, the end of the line. Also, if children are the grown seed of men, the fruit of his loins, they are his harvest and property.

This line of logic leads straight to the fact that even today, in many parts of the world, children belong to the father. A Wodaabe woman living in the African country of Niger can leave her husband, but she'll have to leave the children too because they belong to *him*, not her. Reproductive theory ties women down tighter than a ball and chain. In rural Turkey, according to Delaney, 'because of the meaning of paternity' children belong to the father in the case of divorce, and to the father's parents or relatives in the case of his death. If the widow wants to remarry, she must leave the children behind.[7] This may seem unfair to us but our western culture was, until very recently, much the same. In Britain, women have acquired parental rights bit by bit over this century, with various pieces of legislation, culminating in parity with men only when the Guardianship Act was passed in 1973.

It is perhaps difficult for us now to imagine what it feels like to be the bearer of children, but not the genetic 'mother' as such. A woman in this case can give herself credit for carrying the child for nine months and getting them both through the birth alive and well. She can look forward to the intimacy that binds mother and child through the nursing years, and enjoy the child growing up (if she's lucky). However, these roles are distinct from genetic parenthood and more analogous to the combined role of surrogate womb, wet-nurse and adoptive parent. To her, 'motherhood' means nurturehood and not parenthood – the passing of inheritance, or genetic continuity into the future. She is a baby-making machine, but they are not her babies.

This year it is possible that over a million baby girls will be murdered in India, China and elsewhere because they are reproductively useless. It is not simply that inheritance laws favour sons, or that girls may require a dowry; these facts themselves stem from ideas pertaining to physical inheritance. Why leave your land to a daughter when she'll leave it to her children, the grown seed of a stranger, her husband? Essentially, by leaving land to a daughter, you've worked hard all your life to provide a cosy inheritance for another family's DNA or 'seed'. A son, on the other hand, carries the father's 'seed' which can be supported by the inherited land.

Reproductively, we're on the interface between past and future. We have forgotten our own ignorant past, or are unaware of it, and have no idea where the future is taking us. Genetic engineering will soon be at the stage where we can choose not only the gender of our children, but much else besides. We hope that medical ethics committees around the world will guide us in humane directions, but there is no guarantee. Meanwhile, the less sophisticated technology of ante-natal scanning is being used in India and elsewhere to establish the gender of the child so that, if it's a girl, she can be aborted. This is infanticide, or more accurately femicide, in the modern age.

Where we are reproductively depends on where we are geographically. I happen to live in England where changes in reproductive patterns are happening fast. The first test-tube babies are growing up. Post-menopausal women are giving birth; while young independent women are choosing the fathers of their children from catalogues at the sperm bank. Magazines for heterosexuals, homosexuals and lesbians alike carry advertisements placed by people looking for reproductive partners; love and romance are not involved, but babies are. Even that old bugbear, paternal uncertainty, has been done away with since Cellmark Diagnostics opened its laboratory in Abingdon, Oxfordshire. Paternity can now be established for the sum of £145 plus tax.

But England is not the centre of the universe. In Africa this year hundreds of thousands of young girls will be held down screaming as their clitorises are hacked off with razor blades. No anaesthetic, of course. The reason? To take away the incentive for pre-marital and extra-marital sex so the husband's paternal certainty can be assured. Such measures make sense to people who think a child is either 100 per cent the grown seed of the husband, or 100 per cent the grown seed of some other man. The woman has no reproductive capacity, and no reproductive rights. Consequently, her sexuality from an early age is defined by the reproductive requirements of men.

Meanwhile, on the sunny Trobriand island of Kiriwina in the Solomon Sea, young girls enjoy promiscuous sex because paternity is not an issue; fathers, in the sense that we understand it, simply do not exist. Because children are thought to be the reincarnation of ancestors, they are nobody's 'property' so nobody orders them around and they lead a remarkably independent life. Elsewhere in the world, children will obey their father's every wish because they think they are his grown seed, as much a part of him as his fingers and toes. Even if the father abuses the children, sexually or physically, the mother can't intercede because they are *his* property, not hers.

In almost every town in the world you can buy a copy of *Time* magazine but, reproductively speaking, we do not all live in the same time. There is a wide divide across which we look at each other in mutual incomprehension and horror. By understanding other cultures' reproductive theories we gain a valuable insight into their mores, traditions and life; and by understanding our own reproductive history, we illuminate our own dark past and, we hope, clarify our place in the present.

Part One

Man the seed, woman the incubator

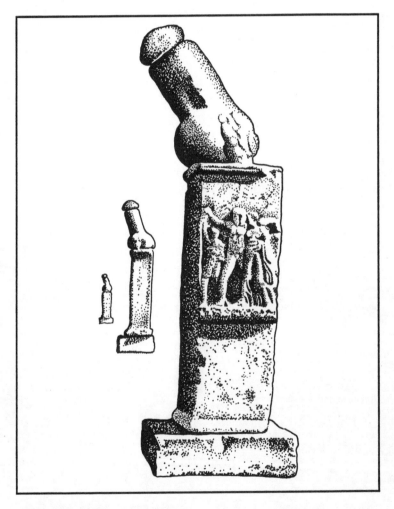

Avenue of Priapus, Delos, Greece
An imaginary reconstruction of the avenue of columns with carved phalluses, island of Delos, 3rd century B.C.

1 Man the seed, woman the incubator

When a man in our culture says of a child 'It's mine', he means, 'It's half mine' because he knows a child gets only half its genetic inheritance from him. When other men say 'It's mine', they mean, 'It's all mine' because they think a child gets all its genetic inheritance or 'seed' from them. There have been many versions of this 'male seed' theory, but only one bottom line: father transmits that which conveys life from generation to generation. Inheritance belongs to men.

Today, we see in the 'male-seed' theory the tail-end of a tradition that has lasted almost five thousand years in places, although many cultures had only a couple of thousand years of it, or even less. The Levite priests who brought the Jewish religion of Jahweh to the people of Canaan most emphatically believed in the male-seed theory, and had to battle hard against other, earlier, theories to get their message across. The male-seed theory pre-dates the prophet Muhammad (c. A.D. 570–632), although in his lifetime matriarchies were still widespread, but Islam incorporated male-seed thinking and the patriarchal ethos became a central tenet of the religion. Sura 2:223 of the Qur'an says this: 'Women are given to you as fields to be sown, so go to them and sow [your seed] as you wish.' Ancient Vedic and Talmudic texts say much the same thing. A Hindu text of A.D. 100 put it this way: 'The woman is considered in law as a field and the man as the grain.' But man wasn't just grain or 'seed', he was farmer, seed, plant and fruit. Children were the fruit of his loins and belonged to him.

Despite the fact that science has provided a different version of reality, the seed-in-soil theory of reproduction persists in many parts of the world today, whether Christian, Jewish, Buddhist, Hindu, Shinto or Islamic. Anthropologist Carol Delaney found it in rural Turkey:

Men supply the seed, which encapsulates the essential
child. A woman provides only the nurturing context for
the fetus. The luxuriant climate of her body is a generalised
medium of nurture, like soil, which any woman can provide.
It affects the physical growth and development of the
fetus, but in no way affects its autonomy or identity. As [a]
villager put it: 'If you plant wheat, you get wheat. If you
plant barley, you get barley. It is the seed which determines
the kind (variety) of plant which will be grown, while the
field nourishes the plant but does not determine the kind'.[1]

When Delaney asked the government-trained midwife about
the role of the ova and ovaries, she mistook the words to mean
vulva and womb and didn't understand their role in conception.
Delaney writes: 'her scientific training seemed to be little more
than a veneer, which quickly peeled away under the impact of
more pervasive cultural notions'.[2]

These notions are revealed by language. A man plants the seed,
tohum, in the female field, the *tarla*. 'To be pregnant', *gebe kalmak*,
translates as 'the seed remained and will swell up'. The male seed
is referred to as *canh*, as 'being alive, as having soul or life in it'.
The male role is 'to put the seed in', *döllemek*, the vagina, *dölyolu*,
literally, 'seedpath', leading to the seedbed, *dölyataği*. The father's
side of the family is the *sulb tarafi*, *sulb* meaning loins, descend-
ants and seed; while the mother's side of the family is the *süt tarafi*,
the milk side.

Man begets; woman gives birth. She is the non-generative
incubator, the seedbed. Her role is to nurture, like soil. These
villagers believe the woman has breasts in her womb from which
the foetus sucks blood; elsewhere, people believe that menstrual
blood goes to make up the matter of the child. All over the world
we can see theories which in diverse ways credit the woman with
providing material or food – that has never been in doubt – but
she wasn't credited with having generative seed.

'Naturally', in these circumstances, inheritance is through the
male line. In Turkey, the words *hane*, *kök* and *sülale* prove that.
Only a man has an *aile*, a family; a woman is just part of one –
her father's, husband's, brother's or son's. A woman is like a leaf
falling to the ground ready to become earth again; she is reproductively transient. A man is the seed that goes on and on; only

he produces *filiz*, shoots or buds. Only man has continuity into the future.

This single generant theory was difficult to refute because seeds do seem to be singular. Western science did not establish plant sexuality until Camerarius came along in 1694, and before then most agreed with Aristotle that plants were single-sexed. When it came to explaining reproduction in those species with two genders, like humans, Aristotle was at a great disadvantage in not knowing about the ovum, the female seed. He thought it was a matter of seed + soil when, in fact, it is about seed + seed and soil. To back up his singular generant theory Aristotle referred to a fish he called 'erythrinus,'[3] which apparently had no male of the species, the females generating on their own. Because 'nature does nothing in vain', according to Aristotle, in those animals with two genders it must be the male that 'always completes the business of generation'.

Aristotle compared men to carpenters who, with their creative force, make an object (the baby) using wood (the woman). He also compared men to artists and women to the material they worked on. (So who would you rather be: Leonardo da Vinci, or a blank canvas and a few pots of paint?) Aristotle wasn't short of analogies: 'The male provides the "form" and the "principle of the movement", the female provides the body, in other words, the material. Compare the coagulation of milk. Here, the milk is the body, and the fig-juice or the rennet contains the principle which causes it to set.' He also put it another way: 'The male emits semen into the female and upon entry of the semen the young animals are "set" and constituted and assume their proper shape.'

The woman's role here is essentially passive: she waits for the creative principle to be delivered into her. Aristotle was in no doubt as to which contribution was the most valuable: 'of course the active elements are always higher on any scale, and more divine.'[4]

The idea that the male contribution was of an essentially higher nature than that of the female was repeated in countless forms over the years. The twelfth-century anatomist Ricardo Anglici said that sperm (by which he meant semen) provides the 'spirit and creative power and form', and Thomas Aquinas in the thirteenth-century declared that men's 'active seminal power' gives form to the 'corporeal matter, which is supplied by the mother'.

The word 'corporeal' means 'of the nature of the animal body as opposed to the spirit; physical; bodily; mortal'. The distinction between men and women was very clear; men provide the blueprint for life, and the impetus for it, as well as the spirit which brings it alive; while women provide the material for men's creative principle to work upon. This fundamental distinction would endure for centuries, unaltered by arguments over whether woman's 'matter' was contributed by her menstrual blood, placental blood, sexual secretions or ovarian material.

Things might have changed in 1672 when Regnier de Graaf, a young surgeon in Delft, Holland, discovered, in a rabbit, what he thought was the ovum. It was in fact the sac which contains the ovum within the ovary, now known as the Graafian follicle. It was not this anatomical inaccuracy, however, which prevented the newly discovered 'egg' from being seen as the generative female seed. Science had for so long now thought of the ovary and whatever it contained as nurturant material that de Graaf's 'egg' could be seen as a detail of this basic scenario, as if the ovary was the shopping bag full of supplies for the foetus, and the 'egg' was a sub-unit, an item of shopping. Also, by a strange quirk of fate, three years later, in the same town, sperm were discovered.

Writing of this discovery in 1853, some 178 years later, the German zoologist Rudolf Leuckart said, 'there is scarcely any discovery in the realm of animal biology which has aroused so general an interest as the discovery of these motile seminal corpuscles'. Embryological historian F.J. Cole says they were talked about everywhere and caused a great stir. No wonder; men thought they had found the singular male seed.

Certainly this is how Antoni van Leeuwenhoek presented his discovery to the world. He said that the 'spermatic animalcule' is an earlier stage of the foetus, in the same way that an insect larva is an earlier stage of the insect it becomes. The sperm, then, is the actual material from which the new life will develop. Each ram sperm, for example, contains a lamb but that sperm will not assume the appearance of a lamb until it has been nourished and grown in the uterus. Leeuwenhoek thought de Graaf's 'egg' provided the nidus for the animalcule, a warm nest with built-in storehouse – a nutritive womb within the womb.

This scenario reflected perfectly the established division of labour: man is the active, creative person while the woman stays home and provides warmth, comfort and dinner. Which is why, perhaps, it was so appealing to the men (and they were all men) of science. Cole makes the other crucial point: 'The widespread support which animalculism received may be partly explained as a result of subconscious bias. Such a view endorsed the superior

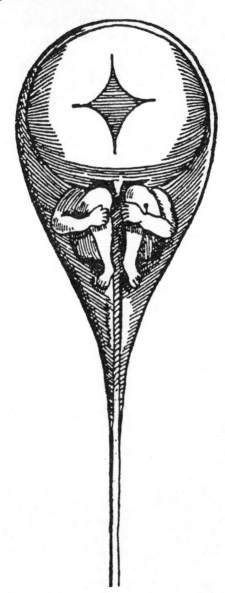

Homunculus
The little pre-formed person in the sperm. An imaginary representation of what a sperm might look like, if able to be seen clearly, drawn by Nicolaus Hartsoeker in *Essai de diotropique*, 1694.

status of the male sex, through which alone the distinctive characters of a species were preserved and inherited.'[5]

Leeuwenhoek was an uneducated haberdasher who happened to be brilliant at making microscopes. He started his hobby as a young man and ground those lenses until he could achieve magnifications from 40 to 270 diameters, using a single lens. His family and neighbours ridiculed him, and the business suffered, but Leeuwenhoek was mesmerised by the tiny world he found when staring down his microscopes and became a man obsessed. Although not trained in science, Leeuwenhoek carefully recorded everything he saw, including bacteria found in stagnant water and weird living things which made up the plaque scraped from teeth. He correctly noted the size of red blood cells relative to bacteria and made important contributions to our understanding of eyes, muscle structure, teeth, and so on. It was all this, as well as the fact that he discovered sperm, that ensured Leeuwenhoek would become known as 'the father of microbiology and protozoology'.

Ironically, Leeuwenhoek's 'animalculism' or sperm theory, which was taken very seriously for 125 years, might never have made it into the annals of science had it not been for de Graaf, the discoverer of the Graafian follicle. De Graaf was a foreign correspondent of the Royal Society of London and was much too much of a gentleman to fail in his duty of bringing Leeuwenhoek's interesting work to the attention of the scientific community. He duly introduced Leeuwenhoek to the Royal Society, and thus to the world. Leeuwenhoek didn't return any favours, however, and de Graaf wasn't loaned, sold or leased any of Leeuwenhoek's brilliant microscopes. But then Leeuwenhoek wouldn't lend, sell or lease his microscopes to anybody, including the eminent gentlemen who now travelled great distances to see him. What he did do was look down one of his microscopes and declare that he had seen the ovum, and it was definitely non-generative. This mistake might have been corrected much earlier had it not been for the fact that, on his death, Leeuwenhoek's microscopes, although he had intended them to be donated to science, were somehow lost or broken. As a consequence, and because nobody was able to replicate their magnifying power for 150 years, Leeuwenhoek's conclusions were taken at face value for a very long time indeed.

Observations of chickens and eggs led to two strong arguments in favour of Leeuwenhoek's 'animalcules' or sperm theory of life. First, the foetus appears in the egg only after it has been fertilised by the cock. And second, an unfertilised egg, when incubated, decomposes; while a fertilised egg cracks open to reveal a chick. In 1721 Bradley described the 'seminal animalcules' as the 'principles of generation or the beginning of man and other animals',[6] and in 1794 Erasmus Darwin, in his own way, agreed:

> In the fowl the unfertilised egg consists only of the yolk and white, which are the food or sustenance of the future chick. The cicatricula (nucleus) is supplied by the cock and is the rudiment of the new animal, but is not a foetus. Hence the embryo is provided by the male, and the oxygen, food and nidus by the female.[7]

That was his theory, at least. Erasmus knew better than most that science had inadequate information to settle the issue and admitted that 'the process of generation is still involved in impenetrable obscurity'. At the time, not a truer word was said. But while science grappled with the central issue, men continued to go about their lives enjoying their elevated status as sole generant. Women, meanwhile, were in reproductive limbo.

If you're a woman, close your eyes, hold your tummy, and think of it as nothing more than a hollow, waiting receptacle. You have no ovum, no DNA, no 'seed'. You are not a source of life; you are merely the temporary resting place of man's growing seed. It is he who has the sense of expansion that comes with parenthood, not you. Children are his future, not yours. You are temporary, not eternal. You stop here.

Because you don't have any generative material, but you do have a womb in which to carry man's generative material to viability, and breasts with which to nurture it, you seem expressly designed to help man fulfil his reproductive potential. A distinct physiological polarity seems evident. One gender is generative – the male; and one gender is nutritive – you. You no longer carry your own baby-making machine around with you, ready to share with the partner of your choice. You are carrying a man's baby-making machine and he thinks you're an appendage to him because 'nature' defined it so. You're not having babies for your own reproductive benefit. You're not a mother, only a 'nidus', a 'field', a 'nurse'. You are man's helpmate, and the world revolves around him.

'God formed her body to belong to man, and to have and to rear children,' said Martin Luther in the sixteenth century, 'Let them bear children till they die of it … that is what they are for.'

Women were born to labour. According to biology historian Maryanne Cline Horowitz, the point had been made in ancient Greece:

> It is implied in Aristotle's view that the female is in the fullest sense of the word a 'laborer'. She passively takes on her task, laboring with her body to fulfill another's design and plan, and consequently her contribution to the product is of a secondary nature. The product of her labor is not hers. The man, on the other hand, does not labor but works. In the craftsman analogy, Aristotle implied that the male is *homo faber*, the maker, who works upon inert matter according to a design, bringing forth a lasting work of art. His soul contributes the form and model of the creation. Out of his creativity is born a line of descendants that will preserve his memory, thus giving him earthly immortality.[8]

We remember Noah, but who was his wife? She wasn't important, not because she didn't help build the ark and feed the animals, but because we didn't descend from her line. Or so we thought.

2 How biology became destiny

Over the years, to support the notion that men were the repro-
ducers, not women, pronouncements were made about the
biological function of certain processes and parts of a woman's
body, including her menstruation, ovaries, sexual secretions,
ovum, vagina and abdominal veins. 'Science', driven since the
ancient Greeks by male-seed ideology, came to a set of erroneous
conclusions that would be used for over two thousand years to
put women down. Science reinforced religious dogma, and religious
dogma reinforced science. For women, defined as an incubator,
there was no escape.

The menses

When the fertilised ovum falls into the womb, it's comprised of
about one hundred cells and is called the blastocyst. It might take
a couple of days exploring the endometrium, the lining of the
womb, before it decides on the optimum site of implantation. Once
decided, the blastocyst and endometrium lock, bound together
by corresponding molecules. If conception has not occurred,
however, the endometrium will shed part of itself and prepare a
new, perfect environment, ready for next month.

Aristotle didn't know about the blastocyst, or even the ovum,
but he did have his logic. He noted that 'semen begins to appear
in males and to be emitted at the same time of life that the
menstrual flow begins'; and that 'in the decline of life the generative
power fails in the one sex and the menstrual discharge in the
other'.[1] From this he concluded that menstruation 'is analogous
in females to the semen in males' (ignoring the fact that one is
related to the calender and the other is related to sex). Because
menstruation stops with conception, Aristotle deduced that it was
'prime matter', which was given 'form' by the male seed: 'the male

semen cooks and shapes menstrual blood into a new human being'. This was not reproductive equality. Aristotle pressed his point: 'It is plain that the female does not contribute semen to the generation of the offspring. For if she had semen she would not have menstrual fluid; but as it is, because she has the latter she has not the former.'[2] Logic made women impotent.

At this point, we have to look at the philosophies Aristotle worked with. He believed that heat 'concocts' matter and that the more heat an animal generates, the further developed it is; and also, that semen and menstrual fluid were both derived from blood. He called these two by the generic term 'sperma' and in terms of sperma, women were imperfect. The reason? The superior heat of men has boiled off, as it were, the raw material, blood, and left behind semen – blood cooked to perfection. The woman's 'sperma', meanwhile, was not so cooked – a fact proved by the observation that it is more profuse and bloody, and was thus imperfect. This biology defined women as beings not fully developed, like young boys not quite there yet, sort of imperfect men. He in fact called woman 'a mutilated male'.[3]

You might think that if heat means development, and girls mature quicker than boys, girls must have more heat. Not so, according to Aristotle's reasoning:

> For females are weaker and colder in nature, and we must look upon the female character as being a sort of natural deficiency. Accordingly, while it is within the mother it develops slowly because of its coldness (for development is concoction, and it is heat that concocts, and what is hotter is easily concocted); but after birth it quickly arrives at maturity and old age on account of its weakness, for all inferior things come sooner to their perfection, and this is true of works of art so it is of what is formed by nature.[4]

Central to Aristotle's understanding of life was his theory of causation which stated that four factors go to make a thing what it is: matter, impetus, form and goal. In human reproduction the child is the goal; the woman is the matter; and men provide the impetus and form – the energy and blueprint. Like all ancient Greeks, Aristotle believed that impetus and form are far higher on the scale of things than matter, and you can be sure that his perception of women suffered. He even wanted the more divine man to be segregated from the physical matter of reproduction, women:

as the first efficient or moving cause, to which belong the
definition and the form, is better and more divine in its
nature than the material on which it works, it is better that
the superior principle should be separated from the inferior.
Therefore, whenever it is possible and so far as it is possible,
the male is separated from the female. For the first principle
of movement, or efficient cause, whereby that which comes
into being is male, is better and more divine than the
material whereby it is female.[5]

The male-seed/female-soil theory of reproduction was well
established by the time Aristotle came along in the 350s B.C. and
applied his logic to the subject, so perhaps it's not surprising that
his deductions should be consistently anti-female. He had pre-
conceived ideas which were not dented by inconsistencies in his
argument. Biology historian Nancy Tuana says: 'the doctrine that
the female sex was inferior to the male was not a premise to be
proved or justified, but was rather an implicit belief underlying
Aristotle's development of biological theory and an axiom upon
which he founded his theory of reproduction'.[6] Certainly, Aristotle
knew where he stood on the matter of sexual equality: 'the
relation of the male to the female is naturally that of the superior
to the inferior – of the ruling to the ruled'.[7] To Aristotle, it was
'naturally' so.

The extrapolations continued. 'The male stands for the effective
and the active, and the female ... for the passive' said Aristotle,
because life is created from inert, passive 'matter' – from women
– and active, form-giving semen – from man. Also, women's
'minds, unlike their bodies, should be kept free from exertion'
because 'children evidently draw on the mother who carries them
in her womb, just as plants draw on the soil'.

On the basis of his incorrect biology, Aristotle defined women
as imperfect, underdeveloped, immature, inferior, passive, less
divine, and consigned them to a life of separation and mental
torpor. I suppose we should be grateful he left us with a soul (unlike
many who came after) although woman's body precluded her from
being able to pass soul on to her children. That privilege was
reserved for men. The problem with all this Aristotelian mumbo-
jumbo, however, is that his biological-philosophical ideas deeply
permeated western culture for 2,000 years. From the sober quietude
of medieval universities to rowdy dinner conversation of learned,

nineteenth-century professors, men referred to Aristotle to justify the traditional male supremacy.

In the Islamic world, the same tradition of thinking continued through Abdullah Ibn Sina (Avicenna), the influential eleventh-century Persian philosopher and physician, who wrote: 'The seed of man is the principle of movement, for there is no doubt that the female semen is like menstrual blood, and menstrual blood is fit to be matter and not the principle of movement.'[8] And the idea continues today in Turkish villages and, no doubt, elsewhere: 'semen or seed is a finer concoction of blood, and semen is the medium through which the divine, eternal element (soul) is transmitted ... women's blood seems to be something additive, thought of more as food'.[9]

To understand menstruation taboo, it's not the food or nutritive explanation that causes problems so much as the idea that the menses are the matter of the potential child, the blood and guts without form. A menstruation is thus the loss of cadaverous flesh, death, and death is fearful, so menstruating women are to be feared, to touch them is to touch death. Menstruating women became *personae non grata*.

The *British Medical Journal* stated in 1878: 'It is an undoubted fact that meat spoils when touched by a menstruating woman.' Six hundred years earlier, the hugely influential Albertus Magnus (1206–80), known as *Doctor Universalis* for his work in science, theology and philosophy (and canonised in 1931), said: 'menstruating women carry with them a poison that could kill an infant in its cradle'.[10] Ignorance about menstruation made women the bearers of death, not life.

The ovaries

Herophilus (260–200 B.C.), the Greek anatomist who lived shortly after Aristotle, is credited with the discovery of the ovaries although they were certainly known of before this. Strabo reported that the ancient Egyptians performed ovariotomy, the removal of the ovaries, on human females; and several Mediterranean peoples living before Herophilus are known to have done ovariotomies on animals. The operation caused sterility and lack of sexual desire; it fattened the sows and gave camels extra strength and endurance.

Western medicine was much influenced by Galen (129–99 A.D.), the Greek anatomist who thought the ovaries were the source of the secretions that flow from a woman during sexual excitement:

> It is poured outside when it has done its service ... this liquid not only stimulates to the sexual act but also gives pleasure and moistens the passageway as it escapes. It manifestly flows from women as they experience the greatest pleasure in coitus, when it is perceptibly shed upon the pudendum; indeed, such an outflow seems to give a certain pleasure even to eunuchs.

Galen thought that this liquid, which he called 'seed', was expelled from the ovaries during sexual excitement, rather as sperm is ejaculated. However, that is where the similarity ends, because according to Galen, the female seed 'contributes nothing to the generation of offspring'. Why? The female is less perfect than the male for one principal reason: 'because she is colder' said Galen.[11] We're back now to the old heat theory, thought up by Empedocles some six hundred years earlier, which stated that the 'hot' elements of fire and air are essentially male, while the 'cold' elements of earth and water are female. Because of her innate coldness, a woman couldn't physically develop to that point of human perfection – the male – thus 'the female must have smaller, less perfect testes, and the semen generated in them must be scantier, colder and wetter (for these things too follow of necessity from the deficient heat)'.[12] Her seed was imperfect, like her, and impotent.

Interestingly, a parallel idea seems to operate in the Turkish village studied by Delaney. The word *meni* means 'semen or sperm' but was sometimes used to mean women's sexual secretions. This was no female generative seed, however:

> Villagers were quick to point out that woman's *meni* does not contain seed *(tohum* or *döl)*, and that it is the *tohum* or *döl* that carries the *can* (soul or life). In other words, the few who used the word *meni* were referring to sexual fluids that both men and women produce, but made it clear that women's fluids were not generative.
>
> As proof, they pointed out that not all women produce sexual fluids and that women can conceive without producing them. Therefore such fluids are irrelevant with

regard to generation, although they may be an accompaniment to pleasure.[13]

As far as Galen was concerned, what he called female 'seed' – the non-generative lubricant which supposedly came from the ovaries – was some kind of all-round biological helpmate. It gave pleasure, which encouraged intercourse and procreation, it provided lubrication and, as he mentions elsewhere, provides nourishment. In later, more puritanical times, nourishment was to become the keyword associated with the ovaries and whatever they contained.

There were other views. Thirteenth-century 'Doctor Universalis' favoured Aristotle's view that menstrual fluid consisted of the 'matter' of the child, which was given form by the male seed. He found another role for what we would call female sexual secretions: 'preparing and enabling matter to receive the action from the operator, that is, man's sperm'.

Anatomists had always had trouble getting hold of cadavers to work on, especially after the Church had its say in the matter, but the ovary presented unique problems. How could anything get from it to the womb? The route was not at all clear, and for good reason. Nobody could see the two-minute drama taking place within a woman's body each month, as a follicle breaks though the surface of the ovary and spills its contents – about a tablespoon of fluid, containing thousands of hormone-producing cells – in the midst of which is the ovum. It hovers there precariously in the deep space of the abdominal cavity, until the fimbrae (at the end of the fallopian tube), alerted by chemicals, usher the ovum into the safety of the fallopian tube. The surface of the ovary looks a bit wounded for a while, but it soon returns to its normal appearance, innocence itself. Who would have guessed that a seed recently popped out of it, and made a huge leap into another piece of anatomy, the fallopian tube? The whole idea is staggering, like reproductive basketball on a tiny scale. Nobody but nature could have dreamed that one up!

Also, looking at a female cadaver, even under a microscope, it is difficult to see how anything can get into the uterus from the fallopian tubes because between them there are powerful tubal sphincter muscles which don't seem to budge. In a live woman, we now know, those muscles are relaxed by progesterone produced by the ovarian follicle and, although it takes the blastocyst a few hours of hard pushing, it gets through.

The ovum

Through the ages 'the ovum' was a term incorrectly applied to various parts of the female anatomy. And when it was referred to, you can be sure that it proved to be non-generative, or generative only of less important parts. In 1740 Astruc was talking about a female 'egg' being responsible for the formation of the placenta and membranes. According to F.J. Cole in *Early Theories of Sexual Generation*, this was a belief common enough at the time.

In 1672, the Dutch surgeon, Regnier de Graaf, came close to finding the ovum when he discovered the follicles which bear his name. Unfortunately de Graaf 'had the fate of Cassandra, to be disbelieved even when he spoke the truth,' said the man who set out to replicate de Graaf's findings some 125 years later. Not that anatomy teacher William Cruikshank fared any better in 1797 Edinburgh. Despite observing 29 rabbit ova, he was not believed, perhaps, because as it says in a biographical reference, he 'gave way to intemperance'. He drank. In any event, 'he must not have been widely believed; otherwise the true conditions would have been established long ago'.[14] So said the man whom history has credited with the discovery of the mammalian ovum in 1827, Karl Ernst von Baer.

Recalling the days of his discovery in his autobiography, von Baer says he was met by *altum silentum*, silence. 'That the older men would not read my paper or, at least, would not let it shake their convictions, I could well imagine. But that the younger participants, also remained silent – that I found disturbing.'[15]

In fact, the very difference between ovum and sperm would be brought to the fore in the drive to prove women's inferiority. Oh, excuse me, difference. In an influential book published in 1889, *The Evolution of Sex*, Sir Patrick Geddes and J. Arthur Thomson outlined their theory: the universe is essentially a dynamic between katabolic and anabolic, active and passive. You only have to take a (superficial) look at the active fast-moving sperm, and the big fat passive ovum to guess which gender is going to be credited with being katabolic, and which anabolic:

> It is generally true that the males are more active, energetic, eager, passionate, and variable; the females more passive, conservative, sluggish, and stable. The males ... the more katabolic organisms, are more variable, and therefore ... are more frequently the leaders in evolutionary progress, while

the more anabolic females tend rather to preserve the constancy and integrity of the species ... The more active males, with a consequently wider range of experience, may have bigger brains and more intelligence; but the females, especially as mothers, have indubitably a larger and more habitual share of the altruistic emotions ... man thinks more, woman feels more.[16]

The man who discovered the X and Y chromosomes in 1902 didn't fully understand their significance or source. He thought Geddes and Thomson's anabolic/katabolic theory was the best around to explain why germ cells develop either into 'passive yolk-laden ova or into minute mobile spermatozoa'. Perhaps he also agreed with Geddes and Thomson's statement that 'What was decided among the prehistoric Protozoa cannot be annulled by Act of Parliament,'[17] and was against the idea of votes for women because biology defined men, not women, as on the cutting edge of history. As recently as 1987 Cardinal Simonis, Primate of the Netherlands, saw fertilisation as justification for the passive role of women in life because the ovum 'waits' for the sperm, the 'dynamic, active, masculine vector of new life'.

These days we know that the sperm are not the sole vectors, carriers, of life, and Cardinal Simonis should know it too. The ovum, the feminine vector of life, arrives in the fallopian tube after a heroic leap through the abdominal cavity – an act which is dynamic and active, not passive. Once there, the ovum or follicular fluid which surrounds it, sends out a chemical attractant, it is believed, which draws sperm towards it. Of the millions of sperm waiting patiently by the cervix, only a few hundred will be activated in this way. In other words, not all sperm are 'dynamic' or 'active', and those that are have been instructed by the ovum to be so. Researchers in America and Israel are wondering if this is the ovum's way of weeding out the weaklings.

The vagina

One thousand five hundred years after Galen's death the vagina was still being drawn in anatomical textbooks as he had described it, as an inverted penis. It hadn't 'popped out', so to speak, with the final phase of development because, as usual, women lacked enough heat to complete the job.

Galen was struck by the similarities between the male and female genital systems: 'Consider first whichever ones you please, turn outward the woman's, turn inward, so to speak, and fold double the man's, and you will find them the same in both in every respect.'[18] The only difference was that the woman's system was internal and the man's external. A woman's genitals, said Galen, 'were formed within her when she was still a foetus, but could not because of the defect in the heat emerge and project on the outside.'[19] The vagina was defined as an under-developed penis, and the woman as an imperfect man.

Anatomy engravers drew the uterus and vagina looking remarkably like a scrotum and penis. The uterus was given an inverted dent in the middle at the top, to make it look like the scrotal sac. The vagina was the penile shaft, and the labia became a larger version of the penile head.

In the nineteenth century, the difference between the vagina and the penis was being used to explain the fundamental difference between men and women, and to explain why genius was a male preserve. The keyword was 'variability', used in terms of men being more variable – outgoing and different from each other – while women were essentially conservative and more or less much the same. Proof came from the fact that there were more male idiots, and the geniuses were all men. Women hovered around mediocrity. With the 'variability' theory, men of science, traumatised by Darwin, were looking to put themselves on the cutting edge of evolution, with women a kind of buffer zone between apes and men.

'One of the most remarkable and suggestive of the laws of variation', wrote W.K. Brooks in *The Law of Heredity* (1883), was that the exclusively male parts were 'very much more variable' than the female parts.[20] You wonder how he arrived at this conclusion. Brooks obviously hadn't worked in a family planning clinic, or he'd know that there is indeed much vaginal variation. Perhaps he didn't mean that penises differ from one another in size and shape more than vaginas do, but that each individual penis can vary its size and shape considerably.

According to Brooks, 'the male element is the originating and the female the perpetuating factor; the ovum is the conservative, the male cell progressive'. Did this affect the brain? You bet:

> if the female organism is the conservative organism, to which is intrusted the keeping of all that has been gained

during the past history of the race, it must follow that the female mind is a storehouse filled with the instincts, habits, intuitions, and laws of conduct which have been gained by past experience. The male organism, on the contrary, being the variable organism, the originating element in the process of evolution, the male mind must have the power of extending experience over new fields, and, by comparison and generalisation, of discovering new laws of nature.[21]

According to biology historian Judith Genova, writing in 1988, although over a hundred years have passed, the same kind of thinking goes on in science today. She writes: 'In the past, men's greater variability was used to explain their superiority. For example, they are more likely than women to exemplify the extremes of intelligence; today, greater laterality is serving the exact same function.'[22] This is to do with brain hemispheric speciali-sation studies which seem to indicate, to some, that women are left-brain dominant, while men are right-brain dominant. Marcel Kinsbourne, one researcher in the field, writes:

> We have seen that the evidence for sex-differential later-alization fails to convince on logical, methodological and empirical grounds ... Why then do reputable investigators persist in ignoring them? Because the study of sex differ-ences is not like the rest of psychology. Under pressure from the gathering momentum of feminism, and perhaps in backlash to it, many investigators seem determined to discover that men and women 'really' are different. It seems that if sex differences do not exist, then they have to be invented.[23]

Thankfully, and increasingly, we now have feminist historians of biology to point out the absurdities that still masquerade as science. Pity our poor female ancestors who did not have that support.

The veins

Galen, the Greek physician who died in A.D. 199, is described as the founder of experimental physiology because his anatomical investigations were full and accurate, the best antiquity produced. But when it came to explaining the difference between men and women, Galen engaged in some 'very creative anatomy', as biology historian Nancy Tuana puts it. This mattered because, as cadavers were increasingly difficult to get hold of, medical science depended on Galen's observational truths for over one thousand five hundred years.

Relying heavily on the argument that men generate more heat than women, Galen needed to explain the source of that heat. Neither Empedocles nor Aristotle had done it. Galen took as his starting point the commonly-held Hippocratic view that female children were created by combining the (non-generative) 'seed' from the left ovary with seed from the left testicle and planting it in the left side of the uterus. Now all he had to do was prove that the left side was impure. Impurity, it was thought, generates less heat and, in this case, 'imperfect seed,' as described by Nancy Tuana:

> Galen explains that the spermatic and ovarian artery and vein going to the right testis and right side of the uterus arise directly from the vena cava and the aorta below the level of the renal vessels and thus carry blood already cleansed by the kidneys. The corresponding veins and arteries of the left testis arise from the renal vessels going to the left kidney and thus carry uncleansed blood. We now know that the veins and arteries of both sides arise from the vena cava and aorta, but this mistake by Galen proved to be a very convenient explanation of the cause of woman's biological inferiority. Holding that they are fed from veins that pass to the kidneys, Galen explains that 'the left testis in the male and the left uterus in the female receive blood still uncleansed, full of residues, watery and serous.'
> In other words, the blood going to the left side is impure. Because it is impure, such blood contains less heat, which will result in its producing imperfect seed. The male seed, however, fed from the pure blood of the vena cava and the aorta, is able to achieve complete development and is thus more perfect than the female seed.[24]

The famous and otherwise fastidious anatomist Andreas Vesalius, who lived in Brussels between 1514 and 1564, continued the tradition of drawing the veins incorrectly, perhaps because it was needed to prop up the biology that made women inferior to men.

The whole

Concluding her essay quoted above from *Feminism and Science*, Nancy Tuana says:

> The accepted belief in woman's inferiority can be shown to have affected, in various scientists, the process of observation, the interpretation of data, and the justification and defense of theory ... The tenacious defense of the belief in the primacy of the male generative powers in the face of growing evidence against it, reveals the deep-rooted nature of this conviction and the emotional valence attached to it.[25]

But sexism did not stop with biology. Even when the First Law of Thermodynamics was formulated by Faraday in 1831, it was used to confirm women's role as the nutritive domestic helpmate. The argument was that as 'thought and physical energy are mutually convertible', and as women's child-bearing role was so energy-depleting, women should not expend energy on mental pursuits because there wouldn't be enough left for making and nurturing babies. Education for women would weaken the race.

Science conspired to make women's destiny that of helpmate to man. From her womb, blood, ovaries, ovum and vagina, a woman's inferiority could be proved and her role defined. There was probably little use in arguing. Her inferiority was proven by the shape of her abdomen, her lack of facial hair, her 'missing five ounces' of brain weight, and even by the fact that she seldom goes bald. There has never been a shortage of theories. Erasmus Darwin attributed women's lack of generative power to the fact that women are smaller than men. Because men are bigger they 'should contribute as much or more towards the reproduction of the species'.[26] As women already did so much in terms of nourishment and warmth, men must contribute something more valuable: life itself. It only seemed fair!

3 A resemblance to mother

You don't have to spend much time at a playground to see that children can have a striking resemblance to their mothers. We know why this is: the genes a child inherits from both its parents might be 'dominant' or 'recessive', passing on, or not, particular characteristics such as blue eyes or red hair. It's the unique combination of genes that makes each person an individual, with some children looking like their mothers, some like their fathers, and some like both parents rolled into one. But before we knew about the genes and the ovum, how could a resemblance between mother and child be explained, given that people thought she didn't transmit generative material?

Farmers have always known that seeds develop differently depending on the ground in which they are planted. The plant reflects the characteristics of the soil. This might sound like a simplistic explanation as to why children resemble their mothers but the eighteenth-century Dutch professor of medicine, Hermann Boerhaave (who was so famous that he received a letter sent by a Chinese mandarin addressed simply 'To the illustrious Boerhaave, physician in Europe') wasn't being much more specific: 'The father communicates the embryo and first rudiments of life, the structure of the body being already determined and assigned in the animalcules of the male semen in all creatures, which yet receives some alteration according to the different species of animal or female from whence it is nourished.'[1]

How this 'alteration' came about was a matter of debate. In 1714 Andry used 'uterine juices' (possibly meaning amniotic fluid) to explain the intermediate character of hybrids:

> Andry says that when the horse is paired with the ass, the progeny, on the embôitement theory, should resemble the father, since the form is determined by the male sex. The actual result, as is well known, is an intermediate, which has points in common with both parents. His explanation

of this is that the juices which the spermatic worm of the horse encounters in the uterus of the ass, being adapted to produce a greater development of the ears than the juices which the same spermatic worm would have found in the uterus of one of its own species, it follows that the ears of the foetus will respond by a more vigorous growth. But the ears of the horse are not capable of *all* the growth to which such a nourishment is adapted, and therefore we should expect them to be longer than those of the horse, but shorter than those of the ass.[2]

Jean Astruc was into moulds. He said in 1740 that the sperm 'introduces itself into the pore or passage of the ovum, it is therein shaped, and, as it were, moulded; whence it is more or less stamped to the likeness of its mother, whose lineaments are impressed by the author of Nature on the hole or passage through which the animal enters the ovum'.[3] F.J. Cole elaborates on the theory:

> If it is a tight fit, the maternal features are the more deeply impressed on the sperm, but if a loose one, less so. Hence the varying degrees of resemblance to the mother. The female sperm is said to be larger than the male, and therefore undergoes greater moulding to the maternal type. This explains why daughters tend to duplicate the mother, and sons the father.[4]

For some unknown reason, a few years later Astruc decided that daughters look like their fathers more often and sons like their mothers. Did this faze him? No, he just reversed the relative size of the male and female sperm, assuming the male sperm was larger, so it was more impressed in the so-called mould. As Cole says: 'By varying the size and contour of the pore (or passage), the mode of inheritance of any parental characters may be readily explained.' Especially when nobody else knows any better.

A very respectable 'scientific' theory was pangenesis, the idea that as blood travels around the body it picks up information in the form of 'gemmules'. These messengers of inheritance end their exhaustive journey in the genital system where they make up the generative male seed, semen, and the non-generative material of women – the menses or whatever. Long eyelashes and even an ear for music could be picked up in this way.

Given that the foetus spends nine months within the body of its mother, and given the wild imagination of generations of theorists, the possibilities for a maternal influence on the form of the child seemed many. Resemblances to mother were attributed to the general – like 'movements of diverse humours' – and to the particular – her womb, her blood, her gemmules, her juices, her mould, her non-generative 'seed', and so on. But if you weren't convinced by any of the physical theories, you could turn to the mind.

In fifth-century Greece, Empedocles thought 'the influence of the maternal imagination upon the embryo was so great that its formation could be guided and interfered with'.[5] Others believed that a child looked like its mother if she had been suddenly frightened during early pregnancy. Over the centuries there was a great deal of talk about 'the force of the mother's imagination' – a nice vague term you could interpret in several ways. Some meant by it that a child resembles its mother if she particularly wanted to make a child when she had intercourse; others meant that she was particularly desirous of the father at that time.

In addition to women's basic physiology, and her mind, women's sexual response could explain the maternal resemblance. Delaney reports that one old male neighbour in Turkey said that if a woman has an orgasm before the man and produces a lot of sexual fluid, the child will resemble her, and if not, the child will resemble the father.

The trouble with all these theories of resemblance is that they contribute by way of confirmation, tying up a loose end, to the theory that men are the actual parents. They also provide a way to blame women for anything that goes wrong with reproduction.

Before the days of sperm counts, semen was analogous with 'seed' and because men produce semen – ejaculate – whether they are barren of sperm or not, women got the blame for being 'barren'. If a woman produced only daughters you could blame it on her imagination, if not her faulty 'nourishment' – as if her soil was too acid or alkaline. In 1685 Peyer blamed the occurrence of monster-children on the force of the mother's imagination or 'the machinations of the Devil, the foetus being deformed in the uterus as a punishment for sin'.[6] And you can guess who got the blame for sinning.

4 The word of God

In the Biblical life-view, men begat and women bore. In Genesis 5 we're told that 'Adam begat a son in his own likeness, after his image; and called his name Seth.' Seth 'begat sons and daughters' including Enos who begat Cainan who begat Mahalaleel who begat Jared who begat Enoch who begat Methuselah who begat Lamech who begat Noah who begat Shem, Ham and Japheth. Like a baton in a relay race, men passed the seed from generation to generation.

Men began to multiply, as we see in Genesis 6:1: 'And it came to pass, when men began to multiply on the face of the earth, and daughters were born unto them, that the sons of God saw the daughters of men that they were fair; and they took them wives of all which they chose.'

Daughters were born unto *men*, while men were sons of *God*. We hear this again in 6:4, where women bore children *to* 'them', i.e. men: 'the sons of God came in unto the daughters of men, and they bare children to them, the same became mighty men which were of old, men of renown'.

The hallowed position of men started with Adam. In Genesis 2:7, 'The Lord God formed man of the dust of the ground and breathed into his nostrils the breath of life; and man became a living soul.' This didn't happen to Eve. She got no 'breath of life', the creative force, and no 'living soul'. On the contrary, she came out of Adam's rib, a 'helpmeet'.

In the Old Testament Book of Job 10:10, God is asked, 'Hast thou not poured me out as milk, and curdled me like cheese?' This refers to the 'milk and rennet' theory which Aristotle put forward in the fourth century B.C., and which is still heard in many parts of the world today. The Jewish Midrash Rabbah (14,19) explains it thus:

> A mother's womb is full of standing blood, which flows therefrom in menstruation. But, at God's will, a drop of whiteness enters and falls into its midst and behold a child

is formed. This is likened unto a bowl of milk; when a drop of rennet falls into it, it congeals and stands; if not it continues as liquid.

Aristotle believed 'the active elements', in this case rennet, were 'more divine', and no doubt others before and after him followed the same reasoning.

In the Qur'an, in the Sura known as 'the believers', Allah describes the creative process thus:

> We created man of an extraction of clay
> then We set him, a drop, in a receptacle secure,
> then We created of the drop a clot
> then We created of the clot a tissue
> then We created of the tissue bones
> then We garmented the bones in flesh;
> thereafter we produced him as another creature.
> So blessed be God, the fairest of creators!

The idea that creation involves putting 'a drop' into 'a receptacle secure' became embryological truth within the Muslim world. In 'Safinat', the Islamic scholar Raghib explains the 'genesis of the embryo':

> Learned men declare that sperm, when placed in the uterus, is first transformed into a small, round ball, while keeping its original white colour. And this lasts for six days. At the centre of this ball then appears a spot of blood. This spot will be the confluence of the souls (*multaqā al arwāh*). When the creation is completed this will be the heart.[1]

In this version of reality, sperm are directly transformed into a baby, like fruit from seed. There is no mention of an ovum or female seed, and nothing about fusion. The ovum is bypassed so that the 'small, round ball' is thought to develop from semen ejaculated into the uterus, that 'receptacle secure'. Embryological development proceeds not from a fertilised ovum, but from the male seed.

In the Muslim village studied by anthropologist Delaney in the early 1980s, people believe 'the essential identity of a child is derived from the father's seed. The mother's role is simply to provide the "material" to swell its substance.' This theory of reproduction lends itself, Delaney believes, to

a distinction between spiritual/essential and material/
supportive, the first male and the second female. However,
it is perhaps more accurate to characterize the differences
as between a generative principle definitive of God and men,
and a nurturant principle exhibited by the earth and women
... nature and nurture are on one side, creativity and spirit
on the other.[2]

Here, as elsewhere in the Muslim world, they say 'the father is
the second God after Allah'. Delaney reports: 'The male role in
procreation was felt to reflect on the finite level God's power in
creating the world. As God is author of the world, so too are men
of children, and upon this their authority rests'.[3] There is no
concept of women being the 'authors' of children because there
is no concept of female seed. Men and women have completely
different roles: 'Life is embodied spirit. Spirit comes from God but
is passed by men via "seed", whereas women, by the nurture they
give, give body to others and are thus more associated with
bodiliness. They are the means through which essential spiritual
being enters the physical world.'[4]

Men's right to obedience from women is enshrined in Islamic
law in Sura IV:38 of the Qur'an, which says 'Men are the managers
of the affairs of women' and states that if women are disloyal or
rebellious they should be admonished and beaten (some translations say 'lightly'). The justification for this is that 'God has
preferred in bounty one of them over the other.' The word
'bounty' refers, I believe, to the (apparent) fact that men were given
reproductive powers, while women were not, and as this theory
is now clearly out of date, some translations of the Qur'an use
'strength' instead of 'bounty'. However, as Abdelwahab Bouhdiba
comments: 'Married life, then, is hierarchized. The Islamic family
was to be essentially male-worshipping.'[5]

Throughout the Muslim, Jewish and Christian worlds, people
look to Adam and Eve for an explanation of the 'nature' of men
and women. Much significance is attached to the fact that Adam
came first and Eve proceeded from him. Eve is fundamentally
secondary. Delaney says: 'This was (also) one of the villagers'
rationales for women's inferior status: whatever or whoever is first
has precedence. This is held to be true no matter what the context;
in this case it implies that woman is not only different but also
"naturally" subordinate to man.'[6]

The Christian Church has spent much time discussing the phrase 'in his likeness', especially when debating the ordination of women. The argument is that priests should be 'in likeness' to Jesus who was descended after 74 generations of men from Adam, son of God.

The genealogy of Jesus is given in Luke 3:23–38. It is a list of 'the son ofs', starting with Jesus '(as was supposed) the son of Joseph'. Joseph was the son of Heli and the list of sons continues backwards in time through Matthat, Levi, Melchi, Janna, Joseph, Mattathias, Amos, Naum, Esli, Nagge, Maath, Mattathias, Semei, Joseph, Juda, Joanna, Rhesa, Zorobabel, Salathiel, Neri, Melchi, Addi, Cosam, Elmodam, Er, Jose, Eliezer, Jorim, Matthat, Levi, Simeon, Juda, Joseph, Jonan, Eliakim, Melea, Menan, Mattatha, Nathan, David, Jesse, Obed, Booz, Salmon, Naasson, Aminadab, Aram, Esrom, Phares, Juda, Jacob, Isaac, Abraham, Thara, Nachor, Saruch, Ragau, Phalec, Heber, Sala, Cainan, Arphaxad, Sem, Noe, Lamech, Mathusala, Enoch, Jared, Maleleel, Cainan, Enos and Seth, until one gets to Adam, the son of God.

If you managed to read that entire list, you'll note that there are no females on it. That's because females were not thought to be the source of any seed; they nurtured the seed, which is quite another thing. When the list says 'the son of', that's what it means – the child has one parent, the named male. The child isn't also 'the son of' a mother somewhere, perhaps just not important enough to mention. She wasn't a parent; she was an incubator.

In Corinthians 11:7–9, we're told that man ought not to cover his head because 'he is in the image and glory of God: but the woman is the glory of the man; For the man is not of the woman; but the woman of the man.' Man is in the image of God because he has the creative principle, seed, and that makes him glorious. Not so woman. Woman was born from man, from Adam's rib, and is proof of his generative power.

In 1986, Monica Furlong from the Movement for the Ordination of Women said one of the reasons given against their ordination is 'the fact that women do not have a penis and therefore cannot represent Jesus ... (You may find the last argument hard to credit, but I promise that this is how it goes. It is more modestly expressed as a rule which makes it difficult to get, so to speak, the point.)[7]

But what is so important about being 'in likeness' to Jesus? What exactly does the expression mean? You can take it in two ways: in terms of masculine physique, the most distinguishing feature of which is the penis; or in terms of creativity, the ability to give life. They are, of course, one and the same thing because the penis is that which delivers male creativity, the life-giving seed.

This idea was held by the Jewish people among whom Jesus grew up, and by the disciples themselves. When Jesus wanted Mary Magdalene to be included among his inner circle, the disciple Simon Peter said in typical unfriendly tone, 'let Mary leave us, for women are not *worthy* of *life*'. When you think about it, this is an extraordinary statement, from our perspective at least, but at the time it was 'natural'. Women were not *worthy* of *life* – women weren't considered good enough by God to confer on them the ability to create.

With the seed went spirit. By having seed, men partook of 'the breath of life' which Adam received in Genesis 2:7. Only man became 'a living soul', not woman. Reproduction was a joint venture between God and his creative agents on earth – men. To men God said: 'I will pour my spirit upon thy seed, and my blessing upon thine offspring' (Isaiah 44:3).

If a child was not conceived or was miscarried it was thought to be because God had withheld his spirit, as we can see from the story of Jacob and Rachel (Genesis 30:2). In another case, Abraham's wife, Sarah, thought her childlessness was punishment for not living in the Holy Land. Then, after living ten years in Palestine and still not conceiving, she blamed herself, although her reason was unclear. Hagar (Abraham's second wife) falsely blamed Sarah's unrighteousness.[8]

Eventually Sarah did have a child, Isaac, who, loved and most precious, became the object of God's most arduous test of Abraham's faith. Abraham was told by God (in Genesis 22:2) to give him up as a burnt offering. This Abraham set off to do (Sarah, of course, not being the actual parent, had no say in the matter). What happened then is told in Genesis 22:9–12: Abraham built an altar, laid the wood, bound Isaac and laid him on the altar but, at the moment Abraham 'stretched forth his hand, and took the knife to slay his son' an angel appeared and said it's OK, you've proved your fear of God, there's no need actually to kill the lad. Abraham then turned around and saw a ram had been caught by the horns in a thicket, and he was duly sacrificed in Isaac's stead. This image, of Abraham with arm raised, ready to kill his longed-for son, is the image most central to the Islamic faith – it says allegiance to God is more important than anything – but it also illustrates the point that children belong to men, not women.

Monotheism reflected perfectly the monogenetic theory of reproduction. If there was a singular male seed, it could be followed back in time, through generations of men to a male god, the original Creator, the ultimate source of seed and spirit. In turn, the legitimacy of monotheism was confirmed by the male-only theory of reproduction. To challenge ideas about reproduction was to challenge ideas about God, and it wouldn't be too difficult to find yourself up on a blasphemy charge.

In the West, we're so bombarded with scientific information we've got used to moving the theological goal posts. Unlike our anti-Darwinian grandfathers, we don't need to believe that God made the world in *exactly* seven days or, at least, we're prepared to accept that one of God's days is a few million years longer than our own. Such broadmindedness is relatively new. In 1927 a young teacher named Scopes was put on trial in America for explaining evolution to his pupils. The prosecutor was Clarence Darrow, a man who ran against Roosevelt for President. What happens now is that scientists write books explaining how science confirms their belief in God, and most of us are not particularly worried about the literal word of God. People have learned to get around its apparent conflict with science, their faith intact.

Not everyone, though, has been through this process, and many still fear the consequences of scientific information, especially when it concerns reproduction. In the Turkish village studied by Delaney the dilemma centred round the library, which neither schoolgirls nor women were usually allowed to visit. Delaney says the fathers and husbands were adamant that women shouldn't learn anything about evolution or genetics, and were aware that reproductive knowledge is a threat to their masculine authority which 'rests upon a specific theory of procreation that is inherent in and supported by Islam'.[9] By forbidding women access to the library, men said they were 'protecting' women from reproductive knowledge, when, of course, it is their own superior position they are protecting.

If God is truth, these men aren't being very godly. They are not only denying women access to information which concerns their own bodies, they're denying women motherhood and spirituality. However, time and science march on. One day these men will have to face up to the fact that they are not the sole parents, nor the sole repositories of spirit. If God is life, then life is within women, as well as men.

5 The ramifications

The male-seed/female-incubator theory of reproduction 'naturally' led people into certain modes of thinking. Of course God was male; of course men wanted sons; of course inheritance was through the male line. But it went deeper than this. A polarity had been established with men at one end – creative and spiritual; and women at the other – nurturing and unspiritual. This caused lines to be drawn in many inappropriate places, both legal and subtle. Moreover, the theory doubled at a stroke men's sense of 'paternity', giving them entire proprietorial rights over children, and an acute desire to assure paternity. Female chastity became all-important. Control became the byword of an era.

Is it a boy?

Of course men wanted sons. Without them the family 'line' would be at an end, and the ancestors won't think much about that. One had a duty to produce sons. The spark of life is eternal only so long as there are boys, as exemplified by the beliefs still held in Turkey:

> In the villagers' theory only men are able to transmit the spark of life, and it is theoretically eternal as long as men continue to reproduce sons to carry it down the generations. From father to son, father to son, this spark is transmitted. The importance of sons is not therefore something separate from the ideology of procreation but an integral part of it, as is the notion of lineage.[1]

When Delaney asked the villagers how many children they had, they'd generally reply with the number of boys. 'Girls in some important sense do not count', writes Delaney. The villagers have an expression: 'A boy is the flame of the line, a girl the ashes of a house.'[2] A boy continues the flame of life, and can pass it on; girls bring the house to an end – they burn it down.

Anthropologist Edgar Gregersen reports that until relatively recently in some areas of the former Yugoslavia, and in some areas of Albania: 'The birth of a son was so stressed that even though a woman might have borne several daughters and in spite of church pressure, a marriage might never take place. A woman who never gave birth to a son would be regarded as worthless.'[3]

She's 'worthless' because she hasn't continued the family 'line', and he's unhappy because he thinks his own tribe has come to a grinding halt. King Henry VIII of England suffered in this way in the sixteenth century. For the sake of a son, Henry murdered several wives and threw the country into religious turmoil. Little did he know that gender is determined by the twenty-third 'accessory' chromosome contributed by men. And little did he expect his daughter Elizabeth to become, some would say, the greatest monarch England ever had.

When some parents ask 'Is it a boy?', they have murder on their minds, but only if the answer is 'No'. Traditionally, infanticide was acceptable and normal in cultures which imposed extremely tough penalties on women who had abortions; the reason being that, with infanticide, you can check the gender first. Nowadays we have pre-natal scanning and some reports suggest as many as 3,000 abortions are being carried out *each day* in India, following this sex-determination procedure.[4]

It is impossible to know how many baby girls are murdered at birth each year in India, as it is not a subject many people like to discuss openly, but as the registered birth ratio of boys to girls is, in some areas, 138 boys to 100 girls (as against the usual 105 to 100), it seems very likely that female infanticide is still practised.[5] In China, according to Chen Muhua, head of the All-China Women's Federation, *millions* of infant girls were murdered during the last decade.[6] A Pakistani friend of mine remembers from his childhood that local villagers used to roll their baby girls down a steep hill. If they lived or died, it was Allah's will.

'Is it a boy?' was, and is, very much a question of life and death. In *The Women's History of the World*, Rosalind Miles tells us that it was and is pandemic:

> From the earliest existence of historical records down to the present day, to be born female in India, China or the Arab states, indeed anywhere between Morocco and Shanghai,

was extremely dangerous. In pre-revolutionary China, childbirth preparations for thousands of years included the provision of a box of ashes next to the birthing bed, to suffocate a girl child as soon as she was born. Throughout India, methods of killing little girls took ingenious new forms in each different place: they were strangled, poisoned, thrown into the sea, exposed in the jungle, fed to sharks as sacrifice to the gods, or drowned in milk with a prayer that they would come again as sons.[7]

To this list you can add suffocation by placing a wet towel over the girl's face, or a sandbag, or giving a double dose of liquid opium. Girls don't stand a chance, even though infanticide has been outlawed in India for over a hundred years.

The preference is for boys. As one Indian put it, 'having a daughter is like watering a neighbour's plant'. In time, she will move on and join another family. Boys stay put and keep the family land together. They look after the old parents, and continue the family name. Girls are superfluous, and in India their dowries are financially crippling to the parents. Hindus believe their soul is released after death only if a son lights the funeral pyre.

This same belief existed in ancient Greece. According to Maryanne Cline Horowitz:

> A common fear of Greeks was that after death the rites would not be performed for them, and that in consequence their souls would wander around restlessly. The happiness of the dead was dependent on the continuity of descendants who would guard and respect the household hearth and ancestral tomb. A female on marriage left her father's hearth for her husband's; fathers sought sons to perpetuate their line.[8]

In Greece, as elsewhere, 'to keep it in the family' meant 'to keep it male':

> The importance of preventing extinction of the family is indicated by the fact that if the only legitimate heir was a daughter, even if she were married she would be brought home, the marriage would be dissolved, and she would then be married to the nearest male relative. The object was to continue the line through her son.[9]

This son would not be hers, so much as the 'nearest male relative's', for reasons we shall shortly see.

Inheritance

In 1776, the writer James Boswell discussed the question of whether to leave his estate to heirs of both sexes, or to male heirs only:

> I had a zealous partiality for heirs male, however remote, which I maintained by arguments which appeared to me to have considerable weight. As, first, the opinion of some distinguished naturalists, that our species is transmitted through males only, the female being all along no more than a nidus, or nurse, as Mother Earth is to plants of every sort; which notion seems to be confirmed by that of scripture – 'He was yet in the loins of his Father when Melchizedek met him (*Heb.* v11.10).'[10]

It isn't odd that Boswell should consider male relatives, 'however remote', in preference to closer female relatives. The females have no seed to pass on, for one thing. With the male seed concept, moreover, the seed that a man inherits from his father is essentially the same as his brother's, his nephew's, his father's or his male cousin's. All their seed stems from the same source: grandfather. Go back a generation, to the paternal great-grandfather, and the seed is still the same and now there are even more distant male relatives to choose your heirs from.

By passing land through a son a person more or less ensures that the property will stay in the family. The son will marry a woman from another family and have children who will be the son's grown seed. If the woman doesn't produce children, which is usually seen as her fault, she can be divorced and replaced or, if polygamy is acceptable in that society, the man can get another wife. Get four, why not? As long as the family has sons, the inheritance is secure.

However, with this male-seed theory of reproduction, the inheritance is not secure if it is passed to a daughter. She will marry a man from another family and have children who would be the grown seed of her husband, a stranger. In effect, by leaving the land to her, you'd just be providing a cosy inheritance for another family. A man didn't work his entire life for that!

As girls were not destined to inherit the land or family business, their sole direction in life became marriage and the production of children. A girl generally moved patrilocally – to her husband's village – where she would provide an extra pair of hands and, it was hoped, sons. Girl-children were a burden because after you'd spent years looking after them they'd move away and spend their productive years with another family. And if, as is common on the Indian sub-continent, she needs a dowry to get a husband, the expense to her parents can be financially crippling.

Boys, by comparison, often bought wealth into the family in the form of the dowry that accompanied the wife. The hoped-for sons will become workers for the fields, warriors for the tribe and insurance for old age. When poverty is rife and social security not yet a Utopian dream, the gender of a child can mean the difference between life and death to the parents because, while boys consolidate and expand wealth and security, girls threaten it.

'Like arrows in the hand of a warrior are the sons of one's youth,' reads Psalm 127: 4–5, 'Happy is the man who has a quiver full of them.' Unhappy, though, was the man without.

Proprietorial rights of the father

With the male-seed theory of reproduction men assumed automatic, unchallenged, proprietorial rights over children. Men would decide whether a newborn baby would live or die. Men had custody in the case of divorce. Father's word was law. In some places, it still is. In Turkey, Delaney was told that a child owes its father respect and obedience because their very life comes from him. Because children are his grown seed they *belong* to him, although these days a court might decide differently.

Also with the male-seed theory of reproduction, children belong to fathers in the same way that an apple belongs to the tree from which it fell. They are intimately related in a way that the apple and earth, or child and mother, are not. But man was not seed alone, he was farmer and fruit as well. Children belonged to him in the same way that a turnip belongs to the farmer who planted it. More so. Children were a part of him in a way no turnip could ever be. They were offshoots, offspring, the very same stuff as he. Whichever way you looked at it, fathers and children went together.

They still do. Villagers in Turkey have a view 'that imagines men as "creators" and "owners" of children, who partake of their essence and are indeed part of themselves'.[11] When explaining why children belonged to the father in the case of divorce, villagers said men were *çocok sahibi*, literally, child-owners. Women are temporary caretakers of children, not parents as such.

Imagine, if your husband died, having to move your children to your ex-husband's parents' house or brother's. You couldn't start a new life with them somewhere else. Their home is not with you, but with the real 'family', the husband's. If you remarried, your children by your ex-husband wouldn't join you and be part of your new family. You would be estranged from them physically as you are removed from them reproductively. Your connection with your children is entirely dependent on your connection with your husband. If you want to divorce him, you will have to 'divorce' the children too. No wonder divorce is not so prevalent in areas where the male-seed theory of reproduction still exists.

In Europe and America children stayed with their fathers in the case of divorce until early this century. In Britain, fathers' proprietorial rights were slowly eroded by consecutive bits of legislation culminating in 1973 with the Guardianship Act, which finally made mothers' rights equal.

Today, remnants of the fathers' proprietorial rights can still be seen in the 'tradition' of him 'giving away' his daughter, the bride. I've never seen a mother give away her daughter at a wedding although, if marriage marks the transition from child to woman, that might be more appropriate. Even if the father is dead, one of his male relatives will do the giving. They are the appropriate choices because, traditionally, the bride was the 'fruit' of their family seed; she was theirs to dispose of. As soon as the bride signs the marriage contract, she loses her 'family' name – the name of her father – and adopts the 'family' name of her husband – his father's name. Women are nameless in a way men are not, and they had little say in their destiny. In *The Women's History of the World*, Rosalind Miles says: 'Throughout the known world both legislation and social custom enshrined the power of a father to marry his daughter where he chose, and to take any steps necessary to ensure that his choice was obeyed.'[12]

Girls today have much more say in the choice of marriage partner, but men are still, to some degree, in charge of the proceedings. In Egypt, brides are merely witness to the marriage contract their father signs on their behalf. In Turkish villages, the marital relation is always formally arranged by the man, and although the bride's presence is now required by law at the official marriage protocol conducted by the imam, in former times her father and another male relative represented her at the ceremony. They had the right to sign her life away.

Child abuse

In *Women, a World Report*, we are told:

> research indicates that as many as one in four families is incestuous. And, in the vast majority of cases – between 80–90 per cent – it is girls being sexually used by their male relatives, usually their fathers ... Extending these figures to the rest of the world implies that as many as 100 million young girls may be being raped by adult men – usually their fathers – often day after day, week after week, year in, year out.[13]

In western culture, abusers perform their evil deeds with guilt and must employ devious psychological tricks to ensure the child's silence. In other cultures, men have the authority to behave in this way because they, and their wives, think the children are 100 per cent the fathers', to dispose of, as John Stuart Mill said, 'at his own will and pleasure'.[14]

Men who think their children are as much a part of them as their fingers or toes, have, with their proprietorial rights, the right to abuse. It is up to the father how he treats his children. He can abuse them sexually, abuse them physically, marry the daughters off to men twice or even three times their age, and kill babies at birth. In ancient Greece, a father had to decide by the fifth day after birth whether the child would live or die. The mother had no say in it. Just as, today, mothers the world over think they have no say in the destiny of their children. Mother can cry, wail and lament, but father can do what he wants.

Child brides

In our life-view, the onset of menstruation equates with the start of fertile life. However, we do not think of menstrual flow as a potential child – the 'matter', as Aristotle described it, the blood and guts that needs man's 'sentient soul' to bring it alive. If you think menstruation is the 'body' of the child, the monthly flow of blood is the flow of cadaverous flesh. It is a stark visual reminder that progeny are being lost. When progeny (of the male gender) are all-important, and girls and women are not, it isn't very serious if the latter die in the attempt to get the former.

The early Christian Church had much the same attitude. The first Archbishop of Canterbury, St Augustine, said in the first century A.D.: 'Any woman who acts in such a way that she cannot give birth to as many children as she is capable of makes herself guilty of that many murders.'[15] As I have said earlier, Martin Luther, the initiator of the Protestant Reformation in the sixteenth century, said of women: 'Let them bear children till they die of it'.[16] These have been common themes throughout male-seed, human history.

In India, the institution of child brides originated in the idea that menstrual blood was the *materia prima* of the embryo, so girls were married well before puberty, at around nine years of age, and moved to the husband's home where he could have regular intercourse before her first menstruation to make sure her 'first fruits' were taken advantage of. According to Webb in the *Pathologica Indica*,[17] 'if an unmarried girl has the menstrual secretion in her father's house, he incurs a guilt equal to the destruction of the foetus'.

According to the colonial British Government Census of India in 1921, 3.2 million child brides had died during the previous year as a result of the sexual activity of their husbands. This is what the medical records say about some girls who survived the early sexual ordeal: 'A: aged 9, day after marriage. Left femur dislocated, pelvis crushed out of shape, Flesh hanging in shreds. B: aged 10 ... almost beyond surgical repair; M: aged about 10 ... has never been able to stand erect since her marriage.'

Early marriage for women is common in Islamic communities. In 1970 Malika Belghiti surveyed women in rural Morocco and found that 50 per cent were married before puberty, and another 37 per cent were married within two years of it.[18] Delaney also found the marriage of girls before puberty is common in rural Turkey.

Chastity

In terms of reproduction, men are significantly disadvantaged. They can't carry the foetus within them, give birth, or feed from the breast. They can't watch a baby emerge from their body and say with parental certainty 'It's mine'. These disadvantages particularly bother a man when he thinks a baby is either 100 per cent his, or 100 per cent some other man's.

To gain access to the means of reproduction, man had to gain control of the baby-making machine: woman. He felt no compunction in doing this because if 'nature' – i.e. God – defined him as the seed, it also defined woman as his reproductive helpmate – that which brings his seed to fruition. Woman was not an independent being with reproductive needs of her own, but an appendage to the generative source: man. Meanwhile, if a man's authority stems from the fact that he can create life, his honour depends on his being able to guarantee that the child is, in fact, his – which means he must control the sexual activity of 'his' woman.

The worst-scenario nightmare for what I call a 'seed-man' was to find 'his' woman impregnated by another man. God appreciated the problem too. As he made Eve from Adam's rib he said, 'Be chaste! Be chaste!' Women's chastity was to become the primary obsession of the male-seed age.

Under the Levite laws of the Hebrews, possibly first written down around 1250 B.C., an adulterous wife could be stoned to death, but if the husband was adulterous, she couldn't even get a divorce. A woman who lost her virginity before marriage was stoned to death, unless she was a daughter of a Levite priest in which case she was burned alive. If a woman was raped while she was married or engaged, she was stoned to death, but if she was a virgin, she was forced to marry the rapist.

Clearly, the experiences and feelings of women were of no concern. These men were interested only in making sure that each man knew exactly which children were *his*. And, being ignorant as well as paranoiac, they weren't sure how 'soiled' a woman got by intercourse so, even though an innocent victim of rape, she was simply disposed of. Better to be safe than sorry.

The problem wasn't simply that the rapist's seed might develop into a baby, or even that they feared the seed might hang around and develop at some later date, the fear was that the rapist's semen had permanently polluted the woman's body. This idea still exists. Ulla-Britt Engelbrektsson found the idea endured in the minds of Turkish migrants in Sweden, whom she studied in the 1970s: 'Foreign seed ... that is seed from any other than a woman's husband contaminates the field forever making the woman permanently defiled.'[19]

In ancient Rome the adulterous wife could be punished by the death sentence while her husband's adultery didn't even count as a crime. The ancient Egyptians chose to burn their adulterous wives, while the Cheyenne Indian might just invite 30 or 40 of his friends to gang-rape her. Almost without exception, the adulterous wife of his-story could be divorced instantly while, by comparison, a British wife could not get a divorce on the grounds of her husband's adultery until 1923. He, of course, could divorce her.

In the later *Bhagavad Gita*, the Aryan hero, Arjuna, speaks of his fear of the undermining of the 'very structure of society' by 'lawlessness' – meaning 'the corruption of women', leading to 'caste mixture'. The worry was that promiscuity, female promiscuity that is, would mean her impregnation by foreign seed. When that impregnation meant the woman carried a child that was 100 per cent the grown seed of someone from another 'caste', the 'corruption' was double what we might think of it today, knowing as we do that the male contributes only 50 per cent of the seed.

The message of Genesis was that despite God's first order to Eve – 'Be chaste! Be chaste!' – she was not to be trusted. In 'Eve's Story of the Fall', retold by Jewish scholar Louis Ginzberg, she admits that after the serpent plucked the fruit of the tree for her she 'opened the gate of Paradise, and he slipped in'.[20] This euphemism for sex explains women's shameful fate:

> Woman covers her hair in token of Eve's having brought sin into the world; she tries to hide her shame; and women precede men in the funeral cortege, because it was woman who brought death into the world ... And because woman extinguished the light of man's soul, she is bidden to kindle the Sabbath light.[21]

So contemptible and lowly are women that Jewish men daily pray: 'Blessed art though, O Lord our God, King of the Universe, that Thou hast not made me a woman.'

The eleventh-century Islamic leader Al-Ghazali, based in Baghdad which was then part of the Persian empire, used Eve to explain why women were destined to suffer eternally:

> As for the distinctive characteristics which God on High has punished women, (the latter is as follows): When Eve ate fruit which He had forbidden to her in the tree of

Paradise, the Lord, be He praised, punished women with eighteen things: 1) menstruation; 2) childbirth; 3) separation from mother and father and marriage to a stranger; 4) pregnancy; 5) not having control over her own person; 6) a lesser share in inheritance; 7) her liability to be divorced and inability to divorce; 8) its being lawful for men to have four wives, but for a woman to have (only) one husband; 9) the fact that she must stay secluded in the house ...

Number 15 really puts things into perspective: 'the fact that merit has one thousand components, (only) one of which is (attributable) to women, while 999 are (attributable) to men'. Al-Ghazali further comments: 'It is a fact that all the trials, misfortunes and woes which befall men come from women.'

When the Crusaders came back from the Holy Land from the eleventh century onwards, they carried chastity belts in their saddle bags. I doubt that their wives were pleased to see them. In the sixteenth century, the Abbé de Brantôme went for a stroll at the local fair where he found 'a dozen contraptions for bridling up women's parts' on sale at the ironmonger's stall. Excavations in Germany have uncovered the fact that some women were actually buried in their chastity belts.

Men's disadvantage in terms of parental certainty is still reflected in laws today. For example, the *1957 Code du Statut Personnel of Morocco*, which states all family law, grants the husband certain rights *vis-à-vis* his wife, the first of which is fidelity. No such right exists for wives.

TV soap operas the world over echo the fears and uncertainties of men. We wait for the dramatic moment when some poor man will discover that he is not the father of the child he thought was his or, conversely, be told he is the father of a child he knew nothing about. Male uncertainty underlies the dynamic between men and women. 'How do I know it's mine?' men say. Ask a man how many children he has and often the reply is one, two or whatever, 'as far as I know'.

Until 'genetic fingerprinting', science relied on testing blood on the A-B-O system which, because it works on exclusion, could only prove non-paternity. If the blood of the child was AO, the mother was A and the 'father' B, then clearly the child didn't get its blood group from this particular man. It must have come from a man of the O group. If the 'father's' blood group was the same as the child, in this example O, that didn't guarantee his paternity. All it indicated was that it was possible the man was related to the child, but as 17 men in 100 belong to the O group, it also meant there were another 16 men in every 100 who could have fathered that child. Between 1975 and 1985 blood-testing methods became increasingly sophisticated until they reached a point where, after subjecting the blood to 32 tests, and by exclusion again, one could say there was a 0.52 per cent chance that there's another man out there in the wide world with the same blood group, and who could have fathered that particular child. This percentage of probability was a major improvement on the 17 per cent probability of a decade before, but it was still only a *maybe*.

We have to bear in mind that these more accurate blood-testing methods were developed only recently and have not contributed in any significant way to alleviating the misery of uncertainty that has aggravated men for millennia.

Fortunately, the answer to men's paternity fears has now been found in 'genetic fingerprinting'. The issue can be settled once and for all. At least, it could be if everyone knew that the facility is easily available, which they do not.[22] Most men with parental doubts just live with them and quite often develop mistrusting and contemptuous relationships with the mother, and a resentful relationship with the child in question.

Clitoridectomy and infibulation

Clitoridectomy is the removal of some or all of the visible clitoris. Infibulation variously describes the removal of the clitoris, labia majora and labia minora, or the sewing up of most of the labia. In the world today, it's estimated that 100 million women have been subjected to some version of the procedures.

The ordeal happens before menstruation begins, often at around seven or eight years of age. The terrified young girl is held down by female relatives, her thighs spread apart and then, without anaesthetic, a razor blade slashes her sexuality to pieces. Her screams cannot drown out the idea in the older women's heads that no man will marry her unless she has this done.

Chastity has to be assured, and while clitoridectomy takes away the incentive for pre- or extra-marital sex (and sex within marriage), sewing her up makes intercourse physiologically difficult. For many women the 'surgery' is not a one-off event, it's something that follows her through life, a constant reminder of her essential

subordination to men's reproductive paranoia. Nawal El Saadawi
tells us:

> In the Sudan there is a veritable army of *dayas* who earn a
> livelihood out of the series of operations performed on
> women, either to excise their external genital organs, or to
> alternately narrow and widen the outer aperture according
> to whether the woman is marrying, divorcing, remarrying,
> having a child or recovering from labour.[23]

Women put themselves through this because they have no reproductive self-esteem. They consider themselves to be machines for
making babies. Men, meanwhile, have an inflated sense of their
own reproductive importance and feel they have the authority
to demand these strict measures of chastity control.

Contrary to popular belief, even in Islamic countries, the Qur'an
does not dictate female excision. Indeed, it would be contrary to
a very important Islamic principle in marriage, that the woman
should be satisfied sexually. In some places circumcision pre-
dates Islam and Christianity by at least seven hundred years.
Today it is common in Egypt, the Sudan, Somalia, Ethiopia,
Kenya, Tanzania, Guinea and Nigeria, but it was known in Europe
as late as the nineteenth century.

Domestic violence

It became very important for the peace of the land that women
knew their position: under the control of men. If men couldn't
control women, they couldn't control their chastity, and they
couldn't be sure whose babies they were having. When it was either
100 per cent yours, or 100 per cent some other man's, this was
all-important.

According to Rosalind Miles in *The Women's History of the
World*, in pre-revolutionary China: 'any man refusing to beat his
wife every night, against the order of his father, could be thrown
into the dungeon of the local magistrate or landowner'.[24] She
couldn't run away because her feet were bound, she could only
hobble. In modern Egypt she wouldn't get much further, according
to Nawal El Saadawi: 'If a husband beats his wife, with reason or
without reason, and drives her to run away from him to her
family's house, he can if he wishes by force of law, under what
is called the *Beit El Ta'a*, send a police escort to drag her back to

him.'[25] Mernissi reports that although a woman in Morocco can now go to court and apply for a divorce from a violent husband, the judges are so unsympathetic that, 'the right to beat his wife is an almost unchecked privilege of the husband'.[26] In Ethiopia, the Hamar men routinely beat their wives because, as one said, 'if you beat her she'll fear you and do the herding and farming'.[27] And what are the men doing meanwhile? They're herding their sisters to protect them from other men!

The fact is, women need protecting from the men they know best. In America, a quarter of violent crime is wife assault,[28] while in Britain, according to Home Office figures, 41 per cent of female murder victims are killed by their present or former spouse, co-habitant or lover (compared to 8 per cent male victims killed by same).[29]

The authorities in Britain have been criticised for taking an unsympathetic approach to female victims of domestic violence. They arrive on the scene and tell a woman with a broken jaw not to get excited. This may account for the fact that only one beaten woman in three will call the police. Or perhaps they know that only 25 per cent of breaches of injunction ever lead to court action. A woman may just thank her lucky stars she didn't get killed – this time.

According to Hamish Sinclair, founder of the organisation 'Change' which he set up to support men trying to change their violent ways: 'Men do not batter through drink or inadequacy, but because they believe they have a right to control their women … battering is just the enforcement end of men's private law, that they are superior over women.'

Polygamy and concubinage

If a man thinks of his children as his seeds grown to fruition inside the incubator-woman, it doesn't much matter who that woman is. To the male-seed mind, the widespread practice of capturing foreign women during war and bringing them home to become wives, concubines or slaves, didn't involve the idea that men would be altering the tribal 'line', or race, with foreign blood. The ancient Hebrews certainly weren't worried about tribal corruption when they captured 32,000 virgins among their spoils of war with the Midianites, as told in Numbers 31: 32–35. And the Sultans of the Ottoman court were succeeded by sons of concubines, as legitimately as sons of wives, by virtue of the fact that the Sultan's seed was legitimate in either case. They were *his* sons, regardless of in whose womb they had grown.

If a man marries or takes as a concubine a woman from another tribe, she will contribute to the fulfilment of his reproductive potential. However, if a women marries a man from another tribe, she will be used to increase the number of *their* offspring. To keep things 'in the family', the seventh-century Islamic *shari'a* ('divine') law states that marriage is permitted between Muslim men and non-Muslim women but not between Muslim women and non-Muslim men. The *shari'a* is still enshrined in many modern secular laws, for example, in Article 29 of the Moroccan *Code du Statut Personnel*.

Seed-man thought of himself as a ram – able to impregnate many females at more or less the same time – and polygamy was seen as the most effective way of proliferating his progeny. Put it another way: he could 'sow his oats' in several fields during the same season. This idea must have appealed to King Solomon because he accumulated 700 wives and 300 concubines.

In 1967, a study of 849 contemporary human societies revealed that 83 per cent allowed men to have more than one wife, fewer than 1 per cent allowed a woman to have more than one husband and 16 per cent insisted on monogamy. Anthropologist Edgar Gregersen has done his own cross-cultural survey and says: 'For a man to have more than one wife was the Old Testament norm and is the ideal, if not the norm, in most societies in the world.'[30] Even though a marriage might not actually be polygamous, the fact that it could be creates its own pressures for the woman, as Juliette Minces explains in *The House of Obedience*:

> For all that the practice is declining, mainly for economic reasons (less than 10 per cent of marriages in most Arab countries are polygamous), the official acceptance of polygamy constitutes a permanent threat to married women. In order to avoid it happening to them, women feel constrained to be exemplary wives in every respect, to please their husbands at all times.[31]

Conjugal rights

Traditionally, 'conjugal rights' were about man's right to reproduce himself. Women had to make themselves available to their husbands for sex. Not to do so was, for Islamic women, a sin:

> The greatest thing in which obedience is imperative is intercourse, which is the goal of marriage. It is the most important thing the man asks of his wife. It is not permissible for her to refuse it except for a legal purpose such as menstruation, sickness or childbirth. For if she does, she commits sin, and her right of clothing, housing and upkeep from her husband becomes invalidated and God's curse will be upon her.[32]

According to Imam Bukhari, the prophet Muhammad said a woman 'who is asked by her husband to join him in bed and refuses to do so is condemned by the angels who hurl anathema on her until the daybreak'.[33]

Delaney found in the Turkish village she studied that a woman does not have the right to refuse sex. There's nothing unusual about this. Until March 1991, women in Britain didn't have the right to refuse sex either – rape within marriage was legal. As one British woman in seven says she has been raped by her husband, there used to be a lot of legal rape going on.[34] Elsewhere, there still is.

Rape

In *Against Our Will*, Susan Brownmiller describes the early days of marriage:

> No quaint formality, bride capture, as it came to be known, was a very real struggle: a male took title to a female, staked a claim to her body, as it were, by an act of violence. Forcible seizure was a perfectly acceptable way – to men – of acquiring women, and it existed in England as late as the fifteenth century.[35]

What 'married' these women to men was not the fact that they had been forcibly removed, but the fact that they had been forcibly raped. A woman's body has been claimed by entry, and possibly impregnated. She'd be no use to another man, even if she wasn't actually pregnant, because she'd been irredeemably 'polluted' by the rapist's semen. From now on she belonged to him. At least, she was no good for the marriage market and was forced to stay with him. Nobody else would have her.

These days women are still terrified of men and with reason. In Los Angeles, one woman in five has been raped; and in London, one woman in six.[36] Statistics show that we are more likely to be raped by someone we know than by a stranger. In a British Home Office report published in 1989, we were told that 39 per cent of reported rapes are perpetrated by a stranger; 30 per cent by former husbands or lovers; and 31 per cent by acquaintances – 'the sort of men who drop in for coffee', as the report puts it.[37] That means that 61 per cent of rapes are inflicted on us by men we thought we could trust. And this is just the tip of the misplaced-trust iceberg because, as research shows, and as we know ourselves, rapes by known men are least likely to be reported because women fear the police and courts will not believe that rape, as opposed to consenting sex, took place. As ever, rape is about control: men thinking they have the right to control women.

Ethnic cleansing

To the male-seed mentality, rape as a weapon of war is ethnic cleansing. The rapist sows his seeds, delivering 'the enemy within'. It's an insidious invasion, the delivery of a time-bomb, guaranteed to confuse the emotions of women and destroy the honour of their men. If one further has the notion that semen irredeemably 'pollutes' the womb, the rapist destroys the purity of the soil in which the next generation will grow. Along with all the other destruction, the rapist causes the destruction of reproductive integrity. If his seed grows and is born, it will be a permanent reminder of the woman's, and her family's humiliation. It will be the rapist's child, not the mother's, and therefore, in time, as the child grows and has an influence in society, the rapist vicariously returns, to take control again.

Spiritual elitism

Men's spiritual elitism was an inevitable consequence of their reproductive elitism. It seemed as if God had given them the spark of life, the creative principle, and that man's reproductive creativity mirrored in some small way the vast creativity of God. Men were perceived as more godly, more favoured, more spiritual and soulful. Women, meanwhile, lacking any creative 'seed', were of a different order.

Men's reproductive disadvantages, especially with regard to paternal certainty, made them paranoid. The Genesis story of Adam and Eve was constructed (from earlier creation myths, as we shall later see) to illustrate the folly of men listening to women. The message was clear: you can't trust a woman, nor should you. Eve's transgressions permanently disbarred women from positions of authority, especially with regard to spiritual matters. As *The Body of Canon Law* states:

> Woman's authority is nil; let her in all things be subject to the rule of man ... And neither can she teach nor be a witness, nor give a guarantee, nor sit in judgement. Adam was beguiled by Eve, not she by him. It is right that the man led into wrongdoing should have her under his direction, so that he may not fail a second time through female levity.

Women, lacking in creativity and therefore having nothing in common with God, and, moreover, not being trustworthy, hardly fitted the role of spiritual leader. In Islam, as in Christianity, the fact that Adam came before Eve (as one would expect the human seed to do) is used also to explain men's leadership over women.

Men's spiritual elitism, however, does not end there. In Islam, hell is populated mostly by women, and heaven, mostly by men. Muhammad himself, in a nocturnal ascension, 'noticed that hell was populated above all by women'.[38] Delaney confirms that this notion lives in the minds of Turkish villagers: 'Hell seems to be populated mainly by women, for it is felt they are morally weaker and more easily seduced into transgression.'[39] Heaven, meanwhile, is a paradise for men, inhabited by 'houris', voluptuous sexual maidens there for the pleasure of the Chosen Ones – men: 'Whenever one sleeps with a houri, one finds her a virgin. Indeed the penis of the Chosen One never slackens. The erection is eternal. To each coitus corresponds a pleasure, a delicious sensation, so incredible in this vile world that if one experienced it one would faint.'[40]

Needless to say, there's no male equivalent of a houri for women to enjoy in *Janna*, the Islamic paradise, presumably because the women are busy burning in hell, *Nar*.

As Delaney found in Turkey, reproductive biology, 'soul', and entry to heaven are inter-related subjects:

The ambiguity about whether women go to heaven seems clearly related to the question of whether women have souls, and what 'soul' means. In some contexts villagers will say that women do have souls; yet in others it is implied that they do not. *Can* is the word they use for soul, and as noted earlier it also means 'life'; since it is *can* that the male transmits via seed, women do not have *can* in the same way that men do. If women do have souls, it is clear that those souls are more embedded in the carnal aspects of life than men's, making it more difficult to extricate women from the physical, material world and release them to heaven.[41]

Here, as elsewhere in the Islamic world, women are debarred from taking a spiritual office, they cannot attend a mosque, they are not blessed before making the pilgrimage to Mecca, they cannot wear special clothes for the pilgrimage, the *ihram*, nor can they spend their time reading the Qur'an when they get home. Women are not allowed to attend burials; and when she dies, a woman (unlike a man), is not blessed at the mosque before burial. The afterlife, the spiritual life, is the domain of men.

Ironically, during the fast of Ramadan, it was the women who kept the fast, many men did not. Delaney wonders why 'women have internalized a view that they are more sinful and therefore need to do whatever they can to help themselves attain heaven. In the hierarchy of being, men are closer to God; representing Him in the family as they do, they may think they do not need to continually prove their worth by such practices.'[42] Men's spiritual elitism may have made them spiritually complacent, as well as spiritually arrogant.

Women, the source of all trouble

Reproduction 'troubles' were barrenness, only producing daughters, and monster births or deformities. All were blamed on women.

Men without sperm, or with a sperm count too low to fertilise an ovum, or sperm lacking motility, still produce semen. Sperm, as we now know, constitute only a small portion of the ejaculate which comes also from the seminal vesicles, two small sacs behind the bladder, and the prostate gland which surrounds the upper urethra. Because infertile men produce the same stuff as fertile men (as far as anyone can tell without the aid of a microscope), barrenness was blamed on women. All men seemed to have 'seed' but women were like soil – very fertile, averagely fertile or barren.

The soil was also blamed if it produced only daughters, as if it were too acid or alkaline to produce sons. Monster births and deformities were blamed on women's thoughts while the baby grew inside the body, or were seen as punishment for her sinfulness. Women, as anyone living under the influence of Judaism, Christianity or Islam knew, were like Eve – basically sinful – while men were not.

Nowhere to run

With the male-seed life-view, women and children became chattels counted alongside men's animals and possessions to be sold or bartered when their menfolk deemed it fit. Men had free will, women did not. While some men were slaves, all women were slaves tied to their masters: fathers and husbands. One grotesque permutation of this relationship continues to this very day in India with the practice of *suttee*. The woman allows herself to be burned alive alongside her husband's cadaver on the funeral pyre. No longer able to serve her husband in life, his family persuade her to serve him in the afterlife.

A son, if badly treated at home, might decide it was worthwhile to stay put because at least he would inherit the land and house. Younger sons could run away to sea or go and seek their fortune in the city. Daughters, however, wouldn't have made it over the local hill.

Women were perpetual prisoners of their gender. Some still are. In the Sudan and Saudi Arabia today a woman cannot book an air passage or book into an hotel without the written permission of her husband or guardian. A woman might be fully grown, with half a dozen children, but even if her husband is dead she must have the permission of a man – if not husband, then brother or eldest son – before she can venture outside the house. If 'a woman's place is in the home', it's because men have imprisoned her there.

6 Women's liberation from incubator status – the story

In the creation debate, the fundamental question was: is the baby a new creation formed out of both male and female material, or is it simply an unfolding, pre-formed seed? – the theories of epigenesis and pre-formation respectively. The male-seed/female-incubator life-view was equally at home with either approach.

Aristotle was an epigenisist in that he believed reproduction was achieved by mixing male semen with female menstrual fluid – but, as we know, he was no feminist. Neither was Galen, who thought male and female 'semen' mixed, but that female 'semen' was sterile. This tradition of belittling the female element is a feature that winds its way through the whole story, starting with the ancients and ending with the desperate nineteenth-century ploy of asserting that the newly-discovered ovum had no nucleus so, unlike the sperm, it couldn't be a cell. The sixteenth-century medical schools of Europe taught that women had 'semen' but declared that it was 'ignoble', or lacking in 'copious spirit', or was 'useless and vitiated'. The verb 'vitiate' means incomplete, imperfect, faulty, impaired or spoilt. Supposedly, only men were perfect enough to procreate.

As a theory, epigenesis was stopped in its tracks in the 1670s by the two discoveries in Delft – de Graaf's 'ovum' (the Graafian follicle, in fact) and Leeuwenhoek's 'spermatic animalcules' (sperm). Science decamped to an either–or position: either the baby was pre-formed in the ovum, or in the sperm. 'Spermism' was the favoured view.

'Pre-formation Doctrine', as it became known, held that the 'seed' (be that sperm or ovum) contains a potential person which (in male or female respectively) carries future generations of seed within it. In other words, all human life was encapsulated at the beginning of time either in the body of Adam, or the body of Eve.

From a feminist perspective, ovism was not as revolutionary as one might at first think. The ovist could certainly argue that within the 'ovum' there was a potential person waiting to get out, but everyone knew that intercourse and emission of semen had to happen first. As life only starts after this event, one had to assume that the semen contributed *something* vital, if not the body itself. Ovism could thus imply that the ovum was a mini still-birth unless touched by the mysterious 'vital impulse' of men, which could be interpreted as life-force, soul or spirit. By comparison, spermism could argue that as life starts with intercourse and semen emission, sperm delivers soul as well as body.

You might think that after the discoveries of the (so-called) 'ovum' and the sperm, someone would put one and one together and come to the conclusion that there was a shared parental contribution. At this time, however, people were too used to thinking in terms of a singular generant. The word 'epigenesis' might convey the illusion of equality, but there was nothing equal about the epigenetic theories that had been around for a couple of thousand years.

Pre-formation

The idea of pre-formed seed was philosophically appealing, as F.J. Cole explains in *Early Theories of Sexual Generation*:

> Religious dogma assumed a mystical and precipitate origin of the world, and assigned to it a dramatic end. All life was created when the world was made, and must face extinction when the original impetus shall have been expended. Everything therefore was accounted for at the beginning, and nothing left for future creation ... To science was allotted the subordinate role of providing the material demonstration of the Divine Plan. The consequence of this may easily be predicted. Any principle which satisfied and expounded the Mosaic convention was acceptable, and the Preformation Doctrine, which allowed that there was no generation in Nature, but only the unfolding of a diminishing series of germs created at the beginning of the world, was adopted almost without question. Proof of it there was none.[1]

Indeed, there were good arguments against Pre-formation Doctrine: caterpillars metamorphose into butterflies and lizards can grow new tails, birds new feathers and crustacea new claws. If they had been pre-formed at Creation, they wouldn't be able to 'create' anew. In 1683 Brunner argued against Pre-formation Doctrine by saying that if, as was generally believed, monsters were the direct result of the mental state of pregnant women, the child could not have been pre-formed at Creation. More scientifically, in 1826, Geoffrey proved that chick monsters could be produced if fertilised chicken eggs were incubated normally for three days and then shaken, perforated or coated with wax so no air could permeate the shell. The conclusion was that if development can be changed in the process, then it was not pre-formed.

The strongest argument against pre-formation was the sheer numbers involved. In a posthumous publication of 1730, Hartsoeker, who thought pre-formation was an absurd idea, tried to calculate how many rabbits were contained in the sperm of the original rabbit. He assumes Creation to be 6,000 years previous, that the rabbit will mate at about six months of age, and estimates that the sperm of a rabbit is at least 10,000 million times smaller than the rabbit itself. He concludes that if a grain of sand is to the whole earth as unity is to unity followed by 60 zeroes, then the original successful rabbit sperm is as unity is to unity followed by more than 100,000 zeroes. Imaginations worked hard trying to figure out the numbers. One could stand with an apple seed in one's hand and imagine a future tree, with more apples and seeds, but go through a few generations and the maths get dizzy. Some found it impossible to believe that all the wheat that had ever been grown had been contained in the original, singular wheat seed.

Others found the numbers exciting. In 1794, Erasmus Darwin thought reproduction was essentially caused by 'a wise superfluity', pointing out that young fish die in their countless millions, and they are more highly developed at the time of their demise than sperm. Then someone pointed out that the fish were not wasted, and neither were plant seeds, because these became the food of other animals; while sperm were just lost. If sperm were so valuable, why did God waste so many millions of them?

This point had been raised by the insect anatomist, Lyonet, in 1742. He also asked, if sperm grow so rapidly in the uterus why don't they grow in the seminal fluid which is their 'native medium'? Turning to the philosophical, Lyonet asked, 'Can we conceive that, in order to form our body, the least noble part of our being, God was willing to create so many hundreds of millions of rational souls only to destroy them?'[2] That raised another point, to which of course there was no answer. The best minds had pondered this question for centuries: when did the soul enter the body or potential body? Indeed, between 322 B.C. and A.D. 1534 they discussed little else.

Spermism raised new questions. In 1717 Goelicke asked why these little tailed beings didn't produce a tailed animal when grown, and pointed out that the theory didn't explain female sterility. In 1722, Maitre-Jan asked how birds' sperm could exist in the seminal fluid without drowning. He thought sperm must have fish-like characteristics, rather than bird ones, and thus actually transform rather than be pre-formed. To some, it seemed unlikely that sperm could be 'seed' because it was so similar in size and appearance when from very different species of animals. Others wanted to know what made sperm so different from the protozoa Leeuwenhoek found lurking in muddy puddles, or the bacteria he scraped off teeth.

Ovism was criticised on theological grounds, as much as anything else. It challenged the order of things as told by Genesis, 2:23: God fashioned Eve out of one of Adam's ribs and Adam said, 'This is now bone of my bone, flesh of my flesh; and shall be called woman, because she was taken out of man.' The implication of ovism, of course, was that, on the contrary, women did not come from man, but man from woman.

At least, that the *body* of the foetus came from women, some were prepared to accept. But they would go no further. Pouchet, writing in 1847 – 175 years after de Graaf and 20 years after von Baer's discovery of the true ovum – fully accepted the ovist position that the ovum contained all the foetal material but noted equally that it couldn't develop until man had delivered his mysterious 'vital impulse', sometimes identified as semen and sometimes as the *aura seminalis*, the pungent aroma of semen.

The strongest scientific argument in favour of ovism was that early signs of visceral development had been observed in unfertilised eggs. Malpighi, Haller and Spallanzani had seen it in chickens and frogs, while Pouchet had seen it in molluscs. Spallanzani, impressed at how a caterpillar egg can become a butterfly, thought frog's eggs contain everything necessary to form new frog bodies. Some were impressed by the fact that a female aphis can reproduce without a male, while others were misled into ovism by observations of dermoid cysts or ovarian pregnancies.

As ever, philosophical considerations came into play, and some became ovists by default. They found the epigenetic idea of a new creation too mysterious to fathom and spermism, which accepted that God should waste so many million precious seed, so incredible as to be impious. Ray, in 1693, thought it was much more likely that God would provide the seed in limited number – in the form of female eggs which 'may if need be, be all brought forth, and come to perfection'. On the other hand, ovism could not explain how the ovum got from the ovary into the womb. Because the ovary is not in direct contact with the ends of the fallopian tubes, and because the sphincter between tube and uterus is so tight, it seemed physiologically impossible.

You can see how much scope there was for argument between the spermists and the ovists and, indeed, the pre-formationists and the epigenesists. As we now know, all this was a waste of time because reproduction is caused by sperm *and* ovum, the pre-formed DNA from *both* mother and father, which *mixes* epigenetically. If the scientists of the seventeenth, eighteenth and nineteenth centuries had had the benefit of our technology they would see in modern reproductive theory aspects of their own. They were all right, but all wrong!

The whole thing was a mess. In 1685, the teacher Drelincourt collected 262 incorrect theories of reproduction which, with his own, made 263. A hundred years later, Senebier, who also had it wrong, wrote this:

> In vain a host of ingenious and profound men have assembled observations on this important matter; in vain the boldest genius has analysed this capital question. The passing of centuries results only in a succession of errors, and whilst darkness the most profound has always obscured from the faculties of physiologists the secrets of generation, ideas the most incredible, contradictions the most flagrant, and travesties of nature the most incoherent have constituted the knowledge of those who flattered themselves on explaining the phenomena.[3]

Until the facts of life could be experimentally proved, the misconceptions and arguments were bound to continue. The conflict only entrenched the male-seed orthodoxy, who could mock the confused scientists and point confidently to the 'Good Book' which explained Creation to their satisfaction. And what went on in the rarified world of science hardly touched the general public for whom life went on as normal – with men in control.

The pre-formation debate was impossible to settle and as the nineteenth century approached, scientific minds turned to a 'new epigenesis' theory. It was from this perspective that J.M. Goode, writing in *A New Cyclopaedia* of 1813, reflected on the enduring and even recent success of spermism: 'The system of generation *ab animalculo maris* was still triumphantly maintained, and the feeble exertions of the few who had sense enough to oppose it was drowned in the multitudinous vociferation of their opponents.'[4]

Clearly, spermism was alive and kicking when we turned into the 1800s, but after a century of scientific inertia, the drive for knowledge was again on. The 'new epigenesis' theory had been gaining support since Wolff suggested it in 1759, spurred on by new discoveries in the laws of magnetic and chemical attraction. So when, in 1824, Provost and Dumas proved that frog eggs are fertilised by the entry of spermatozoa, the scientific world could guess what was coming – an epigenesis involving egg and sperm. The question remained, however, is it the same for mammals? For three years nobody could say and then, in 1827, von Baer discovered the true mammalian ovum and the journey of understanding could begin.

The equalists

In this long history there had been a few brave souls who had the facts of life right – philosophically speaking, at least. Perhaps the first of these was Epicurus who founded his school of philosophy in Athens in 316 B.C., six years after Aristotle's death. The Epicurean doctrine of the two seeds competed for some centuries with the Aristotelian theory that men alone contributed seed; 'real' seed, that is. Which theory was better favoured is well illustrated by the fact that while every schoolchild has heard of the illustrious Aristotle, Epicurus is echoed only in the word 'epicurean', the roots of which are 'one whose chief happiness is in carnal pleasure'. Epicurus probably received the character assassination because of his atheist views on divinity and the afterlife, but his reproductive theory didn't flourish either. Nor, as it happens, did his philosophy of humility in ignorance.

Leonardo da Vinci (1452–1519) was a brighter spark than most, in reproductive matters as in everything else. He wrote about a sexual liaison between an Ethiopian man and an Italian woman which resulted in a child, from whom Leonardo deduced that 'the seed of the mother has power in the embryo equally with that of the father'.[5] Having sorted that, he turned his attention to helicopters.

Some equalists only dared to publish their thoughts anonymously or posthumously. The mathematician René Descartes was one of these pragmatists, only allowing his *De la formation de l'animal* to be published after his death in 1664, 24 years after it had been written. He believed that male and female semen mixed, causing a 'fermentation of particles', the movement of which formed first the heart, and then the other embryonic parts. This was a very radical theory at the time and Descartes knew perfectly well that the Catholic Church could come down heavily on him, as it had recently done on Galileo. In any case, his theory – a form of 'mechanistic epigenesis' – was not successful because it couldn't explain exactly *how* the foetus was formed.

In 1745, Maupertuis anonymously published a paper entitled *Venus Physique*, in which he pointed out that astronomers and chemists had recently proved the law of attraction and asked, 'Why should not a cohesive force, if it exists in Nature, have a role in the formation of animal bodies?'[6] He then put forward the theory that both men and women produce seminal fluid which, when mixed, results in a developing embryo. Almost needless to say, this theory was met with much criticism and was not widely accepted. Indeed, as one critic, Remur, accused (thankfully without revealing Maupertuis' identity, which he knew), the 'attraction' theory smacked of occultism.

One can see why equalists like Descartes and Maupertuis were frightened to come out in the open. To talk about reproduction was to talk about creation and God, and they ran the risk of being charged with blasphemy – a crime punishable by death. And the fact was, everything they said was mere conjecture. The situation wouldn't change for them, or women, until the facts of life could be proved.

The liberation

Women's liberation from incubator status took 133 years – from the start of correct theory in 1827, to experimental proof in 1960.

It all started on or around 1 May 1827, when Karl Ernst von Baer looked down his miscroscope in Koningsberg, Germany, and discovered the mammalian ovum. The moment is recorded in his autobiography:

> Taking a single look through the instrument, I recoiled as if struck by lightning, because I saw a small, well-defined yellow yolk mass. I had to try to relax a while before I could work up enough courage to look again, as I was afraid I had been deluded by a phantom. Is it not strange that a sight which is expected, and indeed hoped for, should be frightening when it eventually materializes?[7]

Strange indeed. In the only book written in the English language which discusses von Baer in any detail, Arthur William Meyer says that after his discovery von Baer became 'distressed, distraught, spiritless and emaciated, and suffered constantly from nostalgia'.[8] Von Baer himself attributed his nervous breakdown to over-work and 'ethical considerations'.

Von Baer attributed his success in finding the mammalian ovum to luck rather than merit, to his 'conviction arrived at during my work with the chick' and, also, his poor eyesight: 'My eyes, which because of their near-sightedness have caused me many an awkward moment in every-day situations, have given me excellent service during anatomical research in that they made small pictures quite clear.'[9]

He also had the good luck to choose as the subject of his enquiry the dog in which, apparently, the ovum is more clearly visible than in some other mammals. Haller, on the other hand, whom von Baer says 'was without doubt the foremost anatomist and physiologist of the eighteenth century', had chosen sheep – 40 of them, in fact. In them, the ovum is harder to identify and Haller, working in 1752 with his student Kuhlmann, had been unable to find an ovum until the seventeenth or, more conclusively, the nineteenth day after mating. It was commonly thought in Haller's time, and indeed in von Baer's, that intercourse caused the release, or formation, of the ovum. Haller thought that, to begin with, a fluid was secreted and directed into the uterus, where it became mucoid and, later, produced the ovum by some process of coagulation. Von Baer noted that 'This was precisely the explanation given to me at Dorpat University as to the genesis of the mammalian and, therefore, also the human embryo'.[10] As late as 1826 this version also appears in the first edition of

Burdach's classic textbook, *Physiology*. So when von Baer confirmed the existence of the mammalian ovum he really challenged the status quo. Haller had been an extraordinarily prolific scientist and, as von Baer comments, 'to "study physiology" in the second half of the eighteenth, and as late as the beginning of the nineteenth century meant in fact to study Haller's works'.[11]

Von Baer's landmark discovery was met, as he records, with silence. His paper, *De Ovi Mammalium et Hominis Genesi Epistola*, sold badly and the publisher declined to take on his next. In September 1828 von Baer attended a conference in Berlin and was surprised when:

> not a single one of the anatomists with whom I became acquainted even so much as mentioned my paper. The paper had been made available to the public as early as the middle of January, and it was highly unlikely that no one had heard of it, especially as I had taken care to have it sent to some of them ... I was too proud or too vain to bring up the subject myself in Berlin. Was my work regarded as a joke or a gross mistake? Or was the subject matter itself of too minor an importance to bother about? I really and truly did not know.[12]

What happened at this conference was a microcosm of what was to happen in the world of science later. To begin with, the participants were mesmerised into silence. They knew perfectly well the historical significance of von Baer's discovery and, indeed, that their own dearly-held theories would have to change. Then, on the last day, one bold delegate came forward and asked the question he and everyone else had on their minds: 'Could you not show us the mammalian ovum in the ovary?'[13] Von Baer willingly agreed and the janitor of the anatomical institute hurried off to prepare his bitch for the 'sacrifice'. 'A rather large number of anatomists' excitedly decamped to the dissection room, where von Baer duly provided the evidence. By the end of the afternoon, wrote von Baer, 'As far as I could make out, everybody seemed convinced, although I am certain that they had been rather doubtful earlier.'[14] And so it was in the macroscopic world of science – a shocked, apparent indifference was followed by slow recognition, but eventual fame.

Today, von Baer is known as 'the father of embryology', not only for his discovery of the mammalian ovum, but for his subsequent work, particularly on germ layers. The work of others could now proceed apace. This isn't to say there weren't hiccups. Eleven years after von Baer's discovery Hausmann was still expanding on the old theory that the ovum coagulated from fluid, and there were those keen to agree with him. This was just one of the many small diversions which would not affect more than temporarily the path of truth which was, from now on, basically straight ahead.

Here, von Baer summarises his contribution within the context of the larger reproduction debate:

> The egg is ... an organized part of the maternal body, which almost always requires the intervention of the male procreative substance in order to develop under propitious conditions, into the corresponding animal according to its type. However, in which way and by what agency the male substance, having reached the egg that earlier was only a part and as such destined to die unless fertilized, now renders this egg capable of developing, that is, capable of becoming independent, is as yet not understood.[15]

It was this, the process of understanding, that would take another 133 years. It would involve dozens of crucial discoveries in the fields of embryology, cytology and genetics – and more names than I have room for here. It was a mystery that unravelled slowly and is, indeed, unravelling still.

The second quarter of the nineteenth century was very productive in terms of embryological research, which was given a tremendous boost by the 1824 observation by Provost and Dumas of frog eggs being fertilised by the entry of sperm, and von Baer's 1827 discovery of the mammalian ovum. By 1847, division had been observed in the fertilised eggs of frogs, newts, fish, starfish, hydrozoa, birds and rabbits, and by 1854 sperm had been seen to enter the corresponding egg in various species but not, as yet, in mammals.

In 1861, Gegenbaur proved that the ovum is unicellular in all vertebrates, and in 1875 Beneden provided the first detailed description of the division of the fertilised mammalian ovum. In 1876, Oskar Hertwig became the first to recognise that the essential factor in fertilisation is the fusion of sperm and ovum *nuclei*, and in 1879 Hermann Fol observed that a *single* sperm caused fertilisation.

Collectively, 'These discoveries did for the study of life much what Kepler had done for astronomy and Galileo for physics', wrote

C.D. Darlington in 1964.[16] But nobody seems to have heard of von Baer, yet alone Provost, Dumas, Hertwig or Fol. Bronowski didn't mention any of them in *The Ascent of Man* – one might cynically ask if that's because they are about the ascent of woman – and radical thinker Buckminster Fuller didn't mention them in his book, *The Critical Path*, although in a 30-page list of important scientific dates he found room to mention the invention of the carpet sweeper.

Had there been such a thing as a female newspaper editor in 1876, she might have carried the headline MALE AND FEMALE THOUGHT TO PLAY EQUAL PART IN GENERATION, in response to Hertwig's crucial observation of egg and sperm nuclei fusing. But the story would have seemed ridiculous when she added, *in sea urchins, at least.*

Running parallel to and interacting with embryological research were equally crucial developments in cytology and what we would later call genetics. In 1838 the German botanist Matthias Schleiden wrote that 'the lower plants all consist of one cell, while the higher ones are composed of (many)', and shortly afterwards his friend, Theodor Schwann, extended the theory to animals. In 1885 Kölliker described sperm as cellular in origin and nature and postulated that hereditary characteristics were transmitted by the cell nucleus. When the general statement of fact commonly called 'cell theory' did come along, the knee-jerk reaction of some scientists was to declare that the ovum wasn't a cell because it didn't have a nucleus. It wasn't until 1861 that the ovum was proved unicellular – in vertebrates, at least. The sperm was confirmed to be a cell in 1885. In 1892 Weismann anticipated the chromosomal basis of inheritance with his 'germ plasm' theory, and soon after the turn of the twentieth century Theodor Boveri proved it.

Things were really hotting up now. In 1891 Henking was the first to see the accessory chromosomes and in 1902 McClung demonstrated that they were the determinants of gender. In 1900 three botanists independently recognised the importance of Mendel's work on the hereditary traits of garden peas, which had been written 35 years previously but dismissed as irrelevant by von Nageli, a distinguished botanist at the time. With the turn of the twentieth century, the front-line in research moved from cellular biology into molecular biology. In 1909 Johannsen coined the term 'gene', which was adopted by Thomas Hunt Morgan and others to describe the sub-units of chromosomes. Working particularly with the fruit fly Drosophila, in 1915 Morgan showed that genes are responsible for identifiable, hereditary traits.

At this stage of the game, the facts of human reproduction could be stated, theoretically at least. We would not acquire our current equation of life, 23 chromosomes from the mother and 23 from the father, until 1956 when Tjio and Levan demonstrated that the normal number of chromosomes in humans is 46.

The experimental proof that human beings reproduce in much the same way as frogs, sea urchins and fruit flies came in 1960 when Dr Landrum Shettles published his slim volume of photomicrographs entitled *Ovum Humanum*. For the first time, there it was for all to see: the sperm entering the ovum, and the subsequent cell division which would lead to a foetus and a baby.

Shettles marks the end of our story because women's liberation from incubator status took a very long time to achieve; at what point a particular person chose to see the emerging truth very much depended on their degree of prejudice against it.

The idea of joint parentage was certainly made possible by von Baer's discovery of the mammalian ovum. Von Baer himself suspected it, but it was still too early – both in terms of science and 'philosophical considerations' – for him to put himself out on a limb: 'Perhaps later we shall yet succeed in proving that the formative relations of the germinal vesicula and the spermatic animalcules correspond. We believe to have found a faint trace of the path to this secret. At present, however, it is still so dark and so little pursued that we do not dare risk proposing it.'[17]

Women, of course, were completely absent from the scientific scene which is why, perhaps, nobody came forward to put the embryological discoveries of the nineteenth century in a feminist context. Why should a man make it the subject of his after-dinner speech at the Royal Society when it was guaranteed to alienate him from his sexist peers? Moreover, the embryological work of von Baer and, later, the evolutionary theory of Darwin, raised questions of such philosophical import to men (and women) that the feminist implications of the emerging science of life would be completely overshadowed.

Von Baer, 'the father of embryology', had pointed out the great similarity in the early appearance of embryos, no matter which species they come from, including humans. He also noted that human embryos look at first like fish, then like reptiles, before taking on their human form, thereby strangely following an evolutionary course. Darwin, of course, caused a furore in 1859 when, in *The Origin of Species*, he asserted that men (and women) were descended from apes. All this seemed to fly in the face of the facts as presented in the Holy Bible, whichever version of human generation one chose to discuss.

In Genesis 1:27, God created man and woman in his *own* image, which made human beings very distinct from the animals who, in 1:26, God had given man dominion over. In the more quoted version, 2:7, 'the Lord God formed man out of the dust of the ground, and breathed into his nostrils the breath of life; and man became a living soul'. This was an honour not accorded to animals (or women) and, again, distanced man from animals (and women). In this version, as we know, Eve was created later, in 2:22, when she was taken from Adam's rib.

Late nineteenth-century embryology was not just about the facts of life. The white men of science were concerned to know where they stood in relation to (a) God, (b) animals, particularly monkeys, (c) white women and (d) black men and women. To discuss the reproductive life of sea urchins and the like was thus a much larger issue than even human reproduction and, indeed, there were a great number of people who maintained that no comparison could be made between animals and humans, or conclusions drawn from animal reproduction experiments. It was all much too unsettling, and many preferred to continue the habit of taking the Bible literally.

Whatever the resistance, science drew us inexorably forward, away from the fundamentals – such as where did the seed come from – and into the minuscule details. We looked at chromosomes, then genes, then DNA, and finally the inspiring double helix. As the jigsaw fell into place, the textbooks were rewritten. But, like all textbooks, they presented the facts, rather than the story behind them. As the important embryological observations were left further in the past, we came to have the impression that, although our ancestors didn't know the precise details of gene transference, they believed as we now implicitly did, that human reproduction was accomplished by the joining of male and female 'seed', which we now called ovum and sperm. In fact, they knew nothing of the sort. Until 1827 they didn't even know there was an ovum. There had been a revolution, but nobody seemed to notice.

A cover-up?

You'd be hard put to find any reference to von Baer outside an encyclopedia. Indeed, the whole subject of the switch in life-view over the past 150 years has dropped down a big black hole. You'd think it never happened.

The cynics among us might immediately point to the fact that there weren't any women working in the field of embryology during the late nineteenth century, and only a few working in cytology in the early twentieth. That meant that relaying the news of women's liberation from incubator status was left to men. Bearing in mind the old truism 'the nature of power is to retain your power', did men consciously or subconsciously take part in a sexist cover-up? Did they surreptitiously edit out from the new textbooks those elements of the story which drew attention to the fundamental change in our view of women's reproductive role?

While I certainly believe that there was an element of sexist editing, which has always been a regular feature of male culture, the 'cover-up' took more than that. Several factors converged to make invisible the shift that had taken place in our perception of women, particularly 'mother'. Because the shift happened in so many small steps covering a century or more, it's difficult to identify a specific time at which the revolution took place. Likewise, for women during this long time it was impossible to experience a revolution as such. The barriers were going quietly down, that was all.

Information had to come from science but during the Victorian era at least, as Cole describes it, 'a gulf was fixed between the secluded specialists who discussed these matters in profound or at least technical language and a public who were educated on moral principles laid down 2,000 or more years ago'.[18] The twentieth century brought better education and mass communication but by that time the orthodoxy had assimilated the concept of the human egg and its fusion with sperm. Women's new role was no longer news.

Embryology involves sex and creation which made it open to theological, philosophical and sexual sensitivities. Of all the branches of science, it was more liable to be left out of discussions at public talks or in drawing-rooms, for example, or in front of children. Somehow, the relevant books got pulled out of general circulation, put on the top shelf perhaps, and from there into obscure medical libraries under the heading 'the history of embryology', where nobody would see them.

Over the years we became less species-centric and less reluctant to draw analogies between animal and human methods of reproduction. Indeed, we got so used to discussing life in terms of the birds and the bees, and drawing conclusions about humans from research into insects, reptiles, other animals and garden peas, that by the time Shettles came along in 1960 and showed us pictures of actual human fertilisation and cell division, we were blasé. We hardly noticed.

But there it was, finally, the proof in black and white for all to see. The sexist stragglers could be offered Shettles' pictures showing the undeniable truth of the matter: human life begins when the nuclei of ovum and sperm meet. We'd come a long way since von Baer, not to mention Aristotle. Finally, women had regained their parenthood.

Part Two

Woman the seed, man the stimulator

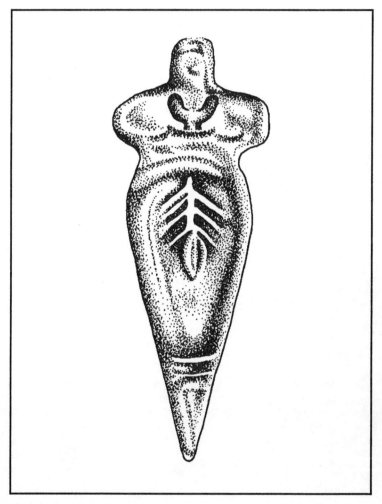

Seed woman, Trento
Bone plate figurine, 5 cm high, Neolithic. Found at Gaban cave, near Trento, northern Italy.

7 Woman the seed, man the stimulator

The earliest evidence we have that people understood men were necessary for reproduction is a stone sculpture in high relief from Çatalhöyük in Turkey, dated betwwen 6000–5500 B.C. Standing at 11.6 cm high, it depicts an embracing couple on the left and a woman holding a child on the right. If this image was meant to depict the notion of 'fatherhood', one might perhaps expect the man to be holding the baby or, if there were any signs in this culture of male authority or supremacy, one might assume the woman was holding the baby on behalf of the man. But Çatalhöyük was not a male-centred society, it was a female-centred one, where the goddess ruled supreme and where, in the words of the archaeologist who excavated the site, James Mellaart, men were 'utterly subject to their mothers'. This was not a male-seed culture.

It is largely on the evidence of this piece of sculpture that many commentators have said 'men discovered their role in reproduction' during the early Neolithic period, around say 8000 B.C. In fact, men discovered their *true* role in reproduction almost ten thousand years later, around A.D. 1900. The word 'discover' should be reserved for events which are accurate. We don't, after all, say that so-and-so discovered the world was flat, because it's confusing; the word 'discover' implies the finding of truth, yet we know the world isn't flat. What we should say of the early Neolithic period is that 'men recognised that reproduction was not an entirely female process and that they had a role in it'. Looking at an image of a couple embracing tells us no more than that, and other cultural clues are required to decipher what, exactly, the male role is.

Intercourse doesn't have to be thought of as the delivery of male seed. We don't think of it that way. The sperm which 'fertilises' the ovum is the smallest cell in the body, at 3 microns across, while the ovum is the largest, at 100 microns. Looking at the two, if forced to say which is the 'seed', I think most of us would say 'the ovum'.

Çatalhöyük group
First known representation of the idea that man was involved in reproduction. High relief carving in stone, 11.6 cm high, dated between 6000–5500 B.C. It was found at Çatalhöyük, near Konya, Turkey. (Museum of Anatolian Civilizations, Ankara, Turkey)

Ten thousand years ago people didn't know about the ovum and sperm. They'd just figured out that intercourse was in some way related to the production of babies which, at the time, was a major step forward in reproduction theory. They also knew that women have 13 menstrual cycles a year, which tells us they probably ovulated, as women do today, 13 times a year. How long an ovum is available to be fertilised each month depends to some degree on the acid/alkaline balance within the woman, and, with that variable taken into account, an ovum may be receptive to sperm for between two and five days each 28-day cycle. That means that over a 365-day year a woman can get pregnant only on between 26 and 65 days a year, leaving between 339 and 300 days when she can't. Add to these days the ones when a woman is breast-feeding, which reduces fertility, and the non-fertile number of days goes up again. Given all this, the connection between intercourse and babies is far from direct and obvious.

Çatalhöyük goddess
Baked clay figurine, 20 cm high, dated around 6000–5500 BC. Found at Çatalhöyük, near Konya, Turkey. The goddess is seated on a throne, flanked by two leopards, and appears to be giving birth. (Museum of Anatolian Civilizations, Ankara, Turkey)

Intercourse does not *inevitably* lead to babies. Indeed, for most of the time it inevitably does not. This fact would have been a strong argument against the idea that the male seed is delivered into the womb with each ejaculation. As we know, male-seed theorists of a later age got around this problem by asserting that non-conception was caused by the woman being too barren or less than fertile. She was given potions and a change of diet to change the condition of her 'soil' and make it more 'receptive'. Without this male-seed logic, however, the fact remains: conception is a mysterious business.

From history and anthropology we have a wide range of reproduction theories which do not involve the idea that semen is the human 'seed'. In the Melanesian Trobriand Islands intercourse is about keeping the baby-passage open and lubricated, nothing more (aside from being a great source of pleasure). Elsewhere, over the years, the essential component of conception has been attributed to the physical action of female orgasm, or to the effects of the movements of lovemaking itself. According to some, the nature of the embrace can affect conception, as anthropologist Colin Turnbull found when he studied the Mbuti pygmies of Zaire who believe pregnancy will not occur if they 'hold each other by the shoulders and not embrace fully'.[1] To them, semen alone is not the determining factor, and if they want a child they 'sleep with the moon'.[2] This indicates a knowledge that women's fertility is associated with a cycle and with the phases of the moon. We'd call it 'ovulation', but what people in other places and at other times called it we do not know, nor do we know if they had the timing right. Even when one knows about ovulation, tracking it for conception and contraception is a notoriously difficult process. However, the fact of moon-related fertility, which has been identified by some people (even if not clearly), could easily become an argument in favour of the idea of a female seed, ripe at a particular time of the month, and ready for the man to 'fertilise' it.

Even when semen *is* seen as the determining factor in conception, it doesn't necessarily have to be perceived as male 'seed', especially when you already think the woman has seed and when nature is abundant with examples of *singular* seed, not two fused ones. And there are alternative explanations for the pro-creative impact of semen. Seeds need watering, so semen could be thought of as water or irrigation for the female seed deep within the woman's body; or semen could be interpreted as fertiliser – like compost – the more of which you put on the soil, the more fertile she becomes; or, from a slightly different perspective, semen can be thought of as essential food for the foetus. With these life-views, although the man is thought of as essential for reproduction, his contribution is non-generative and so he is not 'father' of the child.

A testament to historical reproductive confusion can be found in the notion that women and certain animals can be impregnated by the wind, an idea which existed from 'the earliest times' according to Conway Zirkle:

> Strangely enough most classical and medieval philosophers believed that certain mammals and birds could be impregnated by wind, although they were ignorant of the wind's function in the cross-pollination of the flowering plants. This belief in the existence of what we might call anaemophilous animals is extremely ancient. From the earliest times it existed in Egypt, Greece and the Orient and ultimately it became a part of the intellectual heritage of many races. It spread along the sea routes of the whole eastern hemisphere, and it has been found from Portugal to Japan, and in all of the countries between ... Four types of females were the chief recipients of the winds' attentions; they were mares, vultures, hens and women.[3]

If people can think females can be impregnated by the wind, instead of by a male – stallion, rooster or man – that doesn't exactly credit the male with a crucial role in reproduction.

Looking at the iconography and mythology of the early Neolithic period it seems evident that, in some places at least, when people first made the connection between intercourse and babies they saw the process as the stimulation of female creativity. I think this idea could have started as early as 8000 B.C., that it was the prevailing view until 3000 B.C., that it continued in pockets until as late as A.D. 700, and echoed in our collective psyche until much later than that. A few places were bypassed by the male-seed revolution altogether and there, echoes of our female-centred ancient past can still be found.

When people gave up hunting and gathering and started farming, they began to think in terms of a human 'seed'. As seeds are singular, they were probably only looking for a singular human seed, not two that fuse. As the fruit of the seed emerges from the body of a woman, it would make more sense to expect to find the seed in the body of a woman, rather as one finds the seed of an apple tree coming from the tree itself. They could have thought women produced two kinds of fruit: one male and non-generative and one female and generative, able to reproduce through another generation.

Woman can be seen as the soil, but with the seed already in it. (Rather, in fact, as the uterus is the soil in which the ovum, already inside the woman, grows.) Thinking in these terms, semen could be seen as irrigation. Given enough 'watering', the seed within the body of woman would sprout.

Woman the seed

Consider the fact that of the tens of thousands of prehistoric human figurines discovered by archaeologists so far, only 2 to 3 per cent clearly represent the male. Pre his-story was her-story and, as we shall later see, the goddess ruled supreme, women were in positions of power, inheritance was through the female line and parenthood belonged to mothers alone.

Even without the use of words, the concept of woman-the-seed was powerfully conveyed by visual means. Female figurines are decorated with trees, branches or other vegetation firmly placed upon their abdomens. In some, the seed-shape centre of a plant looks like a womb. When Sir John Marshall excavated Harappa, a city built by the Indus Valley civilisation, he discovered a seal (dated at 2000 B.C.) on which the Earth-goddess is shown with a plant coming out of her vagina. It looks like generativity and femininity were, once, one.

The iconography of prehistory abounds with images of the egg. Eggs appear on vessels of all sorts, often coloured red inside the body of a bird. There were small egg-shaped burial jars and, in Malta, Sicily, Sardinia and other central Mediterranean areas, egg-shaped tombs were hewn out of solid rock. According to archaeologist Marija Gimbutas: 'prehistoric art reveals that the egg in the Old European belief system stands for becoming, regeneration and recreation'.[4] And to comprehend the centrality of the egg in the human life-view, one has only to think about the innumerable myths which state that a great Cosmic Egg gave birth to the world.

Only females lay eggs. Although human females don't lay eggs they, and other mammals, carry the foetus within a membranous sack not unlike the membrane on the inner side of an egg shell. The connection between women and birds, the most common egg-layers, was forcefully portrayed in 'bird goddess' figurines and jugs et cetera, which show the head of the bird in some way combined with the body of a woman.

Man the stimulator

In the history of the world, gods have their own evolution. They start out as the sons of parthenogenetic goddesses (women who reproduce on their own), then become their lovers, then consorts and, in time again, they become deputies to the goddess, exercising power on her behalf. Later still, the myths tells us how gods usurped power from the goddesses, and we begin to see gods as the ultimate creative force.

The early emphasis on the woman is reflected in the iconography our ancestors left behind. Basically, the male is conspicuous by his absence. Instead, we see images of women, 'fertility goddesses' as they are generally called. In time, the male makes his entry, but he does so not as the seed, but as the *waterer*. Gerda Lerner has surveyed the symbols of early civilisations and observes:

> At first, the tree of life, with its fruit – the cassia, the pomegranate, the date, the apple – was associated with fertility-goddesses. At the time of the development of kingship, kings assume some of the services to the goddess and with them some of her power, and have themselves depicted with symbols associated with her. They carry the water-of-life jug; they water the tree of life. It is most likely that this development coincided with the change in the concept of the fertility goddess: namely, that she must have a male consort to initiate her fertility. The king of the Sacred Marriage becomes the king 'watering' the tree of life.[5]

In early Sumeria the idea that men irrigated women with their sexual fluid – stimulating the seeds so they sprang into life – was encapsulated by the myth of the god, Enki. He stimulated a succession of vegetation-goddesses with the gushing of fresh water from his penis. Sumer was the land between and around the southern end of the Tigris and Euphrates rivers, in present-day Iraq. Sumer had 14 major cities, including Ur, and existed from 5000 to 1800 B.C. The land receives little rainfall so, to take advantage of the rich alluvial soil left by the annual flood, the Sumerians invented irrigation. It's thought they did this in Eridu which, dating from 5000 B.C., was perhaps the oldest city of all.

The god at the temple in Eridu was Enki who, with his gushing water, fertilised the most ancient Sumerian deity, Ninhursag, the goddess of the earth.

The god Damuzi (later, Tammuz), the most famous consort of the goddess Inanna (later, Ishtar) was known as 'the Quickener of the Child in the Womb'. The child is already in the womb and Damuzi hurries it along. Aside from meaning 'to hasten or accelerate', the verb 'to quicken' means 'to give, add or restore to a person or thing', and we can see Damuzi as adding that which restores life to the seed within. The word also means 'to animate, stimulate, excite or inspire', and while Damuzi was certainly venerated for doing these things, he was not venerated above the female. Of his relationship to Ishtar, the scholar E.O. James wrote:

> In this alliance she was the dominant partner ... for when he was brought into close connection with Ishtar, in the Tammuz myth, he was her son as well as her husband and brother, and always subordinate to her as the Young-god ... in the last analysis Inanna/Ishtar, not Damuzi/Tammuz was the ultimate source of life and regeneration, though the young god as her agent was instrumental in the process.[6]

In these early days men were not, as they so emphatically became later, carriers of seed, but they were instead associated with life-giving water. The extremely ancient goddess Tara, known under different names from Ireland to Tibet (Taranis, Turan, Terah, Terra Mater and so on), was fertilised by rain from the gods, such as Jupiter Pluvius.[7] In Wales and Ireland, Tara was fertilised by Taran and Torann respectively, their names both meaning 'thunder', perhaps an oblique reference to the thunderous, explosive nature of ejaculation which, like thunder, is followed by wetness – from the sky in the form of rain and from the man in the form of semen. Wetness was also implied by the net-motif, snake or meander decoration sometimes found on phalli discovered in south-eastern European Neolithic or Copper Age sites.

Wetness was associated with men, and the earth was associated with women, but where was the seed? – Was it in the earth, like perennial seeds in the garden? – a concept more likely before the practice of *sowing* seed, which came along only with the agricultural revolution. Even when sowing had become a way of life, it did not immediately lead to the concept of the *delivery* of seed, male seed. It would take another revolution before reproductive power was appropriated by men. In the meantime, men were often wet, like the Cretan sea-god Poseidon, consort to the earth-goddess, who lived not only in the sea, but in springs and sub-terranean water supplies as well. Even that late macho hero, the Mesopotamian King Sargon (c. 2350 B.C.) says, in an autobio-graphical account of his life, that his mother gave birth to him in secret, put him in a basket of rushes sealed with bitumen, and sent him down the river where he was found by Akku, 'the irrigator'. Akku made Sargon 'a gardener' – a euphemism for fructifier of the goddess. This is what men did: they stimulated the female into producing. Men made women fertile, but they were not themselves fertile. That's why families didn't reckon their descent through men.

With the female as the ultimate source of life, matrilineal descent would be the 'natural' order of things, which it was. Despite all the evidence of gods and kingship during this early phase of human history, the importance of the male should not be over-estimated. In *Religion of the Semites*, W. Robertson Smith tells us that the existence of gods did not denote the concept of 'fatherhood', and the parent was the female:

> the history of the family render[s] it in the highest degree improbable that the physical kinship between the god and his worshippers, of which traces are found all over the Semitic area, was originally conceived as fatherhood. It was the mother's, not the father's blood which formed the original bond of kinship among the Semites as among other early people and in this stage of society, if the tribal deity was thought of as the parent of the stock, a goddess, not a god, would necessarily have been the object of worship.[8]

And she was. Since Robertson Smith wrote this in 1894, archae-ologists, digging all over the world, have found tens of thousands of prehistoric figurines in the shape of humans. In south-east Europe alone, 30,000 figurines have been uncovered from 3,000 sites. Given his understanding of the relative importance of women in early society, Robertson Smith would not have been the slightest bit surprised by the fact that only 2–3 per cent of the figurines clearly represent the male.

Animal husbandry

Animal husbandry, in itself, tells us no more about human repro-duction than human intercourse, and people tend to project on to their animals life-views gained elsewhere. For example, when the anthropologist Malinowski asked the Trobriand Islanders pertinent questions about pig-breeding, he found some interesting logic. They had three kinds of pig: the domesticated – which was a delicacy; bush-pig – the eating of which was taboo; and the imported – which cost between five and ten domesticated pigs. So, inter-breeding was certainly a matter of concern. At least, we might think so. The Trobrianders, however, start from the position that 'The female pig breeds by itself',[9] meaning that she breeds without the aid of a hog and, also, without pig-spirit being involved. Consequently, inter-breeding was of no concern and the Trobrianders allowed their domesticated pigs to wander into the bush, where they mated with wild pigs. When Malinowski suggested that they were actually eating the results of inter-breeding (taboo pig), the Trobrianders simply didn't know what he was talking about. One native told him: 'From all male pigs we cut off the testes [to fatten them]. They copulate not. Yet the females bring forth.' As Malinowski notes: 'Thus he ignored the possible misconduct of the bush-pigs and adduced the castration of domestic hogs as final proof that intercourse has nothing to do with breeding.'[10]

Elsewhere, the observation that no ram = no lamb was eventually made but, starting from the position that creativity was essentially female, they would be more likely to perceive that connection in terms of stimulation of female generativity, rather than the delivery of male seed. The difference between the two life-views is, after all, only a matter of attitude and the attitude at the time gave favourable emphasis to the side of the female.

Probably the most famous example of animal husbandry from antiquity is the story of Jacob from Genesis 30: 35–42. Jacob (later renamed Israel by God) had been working for Laban for 14 years and wanted payment. It was agreed that he should take the animals that were striped, speckled and spotted white. Jacob now set about manipulating the stock. He stripped part of the bark of some hazel, green poplar and chestnut branches, revealing some of the white beneath, and placed them in the watering troughs. The purpose of this was that 'they should conceive when they came to drink; And the flocks conceived before the rods, and brought forth cattle ringstraked, speckled, and spotted'. I don't interpret the word 'conceived' in this context to mean the actual drinking of the water made the animals pregnant, as some have suggested;[11] I believe it means that the females were mounted while they stood in front of the watering troughs. But, whichever, Jacob's 'husbandry' involved the fact that 'whensoever the stronger cattle did conceive, that Jacob laid the rods before the eyes of the cattle in the gutters, that they might conceive among the rods; But when the cattle were feeble, he put them not in: so the feebler were Laban's and the stronger Jacobs'.

As a committed patriarch, I believe Jacob would have been thinking in terms of male seed. Presumably he thought that if, at the moment the male seed entered the body of the cow, the image being received by her eyes was spotted or striped white, it would somehow transfer to her offspring, thus making them spotted and striped white. Whatever was going through Jacob's mind, however, this passage proves that erroneous ideas and silly practices were going on six or seven thousand years *after* the commencement of animal husbandry. It is not therefore unreasonable to suggest that during the early Neolithic, and even later, people might have thought that male animals, like men, stimulated the germinal seed in the female animal. And if they did, there was nothing to prove them wrong. As we have seen, when it comes to ideas about reproduction, 'logic' can be bent to prove any point of view.

The blood-line

If women were the source of seed, you'd expect descent and inheritance to pass through them. It did. The original 'blood-line' ran from mother to daughter; men were not involved. Descent and inheritance through the mother-daughter line was the practice in Ghana, Nubia, Libya, Ethiopia, Egypt, Palestine, Mesopotamia, Thrace, Anatolia, Persia, Crete, Etruria, India, Polynesia and, indeed, wherever there are records old enough to prove the point.

Some of our information comes from the ancient travellers. In the fifth century B.C., Herodotus wrote: 'Ask a Lycian who he is and he answers by giving his own name, that of his mother and so on in the female line.' Lycia was a nation-state on the southwestern coast of Turkey and, of the Lycians, Heraclides Ponticus said: 'From days of old they have been ruled by women.' Nicholas of Damascus said of them: 'They name themselves after their mothers and their possessions pass by inheritance to the daughters instead of the sons.' In the second century B.C., Polybius wrote of the Locrians: '[they] themselves have assured me that all nobility of ancestry among them is derived from women and not from men'. In the first century B.C., Diodorus Siculus wrote: 'only the daughters inherit in Egypt'; and as late as the first century A.D., the geographer Strabo said that among the Cantabrians inheritance was through the daughters who 'had the obligation to supply their brothers with dowries'. Strabo also reported that there was no such thing as an illegitimate child in Armenia because children took their names from their mothers.

Academics confirm these earlier reports. Dr Margaret Murray has proved that during most periods the Egyptian royal line passed through the daughters. Men came to power only by marrying the queen or her heiress. Professor Cyrus Gordon wrote: 'In family life, women had a peculiarly important position for inheritance passed through the mother rather than through the father.' Sir James Frazer confirms that: 'In Egypt, the archaic system of mother-kin, with its preference for women over men in matters of property and inheritance, lasted down to Roman times.' One of the original excavators of the pyramids, Sir William Flinders Petrie, wrote: 'In Egypt all property went in the female line, the woman was the mistress of the house; and in early times she is represented as having entire control of herself and the place.' In *Biblical Archaeologist*, Barbara Lesco notes:

> The importance of the mother in the Egyptian family is reflected in the literature of all periods as well as in the fact that Egyptian men, even those of the highest social class and in highest ranks of the civil service or the military, often placed only their mother's names on their monuments and other documents.[12]

When the European invaders came to do 'land deals' with the native American Indians they were staggered to find that it was the women who came forward to sign the documents. Here, in the words of an old Nootka woman from Vancouver Island, Canada, is an account of how things were, and what happened to change things, as recorded by Anne Cameron in *Daughters of Copper Woman*:

> 'My grandmother was the Queen Mother,' she said suddenly and flatly, replacing her cup and picking up her knitting again. 'Her son was the king. She wasn't Queen Mother because he was the king, like in England. He was king only because she was Queen Mother. His son wouldn't inherit

to be king. The Queen Mother's oldest girl, my mother, would become Queen Mother and *her* son – my brother – would have been king. I would have been Queen Mother after my mother because I was the oldest girl, and my son would have been king. Your mother, Ki-Ki, would have been Queen Mother, and if you'd had a brother, he'd have been King. Then you'd have been Queen Mother and your son would have been king.'

'Only it all got buggered', she amended calmly, flashing a funny twisted smile at Suzy and me. 'Got real buggered because the ships came back and instead of chasin' them away because of what had happened before, the people hoped this time things'd be different.'[13]

Charles Seltman explains how things were elsewhere: 'Among the Mediterraneans, as a general rule society was built around the woman, even on the highest levels where descent was in the female line. A man became *king* or *chieftain* only by a formal marriage and his daughter, not his son, succeeded so that the next chieftain was the youth who married his daughter.'[14]

We have many examples of this from history. King Cyrus of Persia was just a commoner who became king only when he married a princess – in this case, Astygates. Indeed, the fairy tale about the princess locked up in the castle's turret by her mean father or brothers should be seen in this light; they are keeping her away from the valiant knight on the proverbial white charger because once she is 'rescued', the father and brothers have lost their power, which went with her. They know that once the knight has consummated his claim, 'married' the princess, he will have the lot!

The phenomenon of royal incest should also be understood in terms of matrilineal descent. Kings held office only because they were married to the source of power, a woman, who was often their sister. When she died, the king was out of a job unless he managed to arrange his marriage to his daughter – that is, the queen's daughter and the rightful heir.

The interface period

Reproduction involves one's parentage, progeny, inheritance, status, spiritual understanding and security. Human beings have discussed it in terms of the human body – male and female – and in terms of gods, goddesses, ancestral spirits, animals spirits and spirits of the forest. Reproduction is a very broad subject, as well as the most profound. Because the facts of life are by no means clear, and couldn't be proved for such a long time, many theories were bound to appear. What is the least likely, given the vast span and scope of human society, is that human beings should have had but one theory: the male-seed view that dominated the 'civilised' world for the last two to five thousand years. That view formed the foundation of our traditions and defined what we thought was the 'nature' of men and women, and it is from this tradition that we project on to the past our own limited perceptions of human beings. Shackled in this way, it's difficult to imagine a time when, on the basis of their reproductive superiority, women were thought 'naturally' superior.

In time, this idea came to be challenged. There was what I call an 'interface period' when some people asserted men were the primary force in reproduction, and others maintained women were. To say this was a dynamic time in history would be an understatement, as we shall later see. Indeed, many wars were fought because of this crucial issue: should babies 'belong' to men or to women? There was a lot at stake. One didn't just give up a theory, one gave up one's children and an entire way of life. Conflict ensued as the two theories existed almost side by side; tribe by tribe, valley by valley, nation by nation.

Male reproductive jealousy was alleviated in some cases by apportioning men a role in terms of 'spirit' – the wedge that often led to their biological takeover. There was also a great deal of 'hedging one's bets', incorporating both theories into one confusing concept. There's nothing unusual about this, people do it all the time. The Irabo of Nigeria continued to worship nature spirits for at least a hundred years after they had been converted to Christianity, and the Trobrianders continue to arrange their lifestyle around the idea of ancestral spirits, and have promiscuous sex and *then* go off to church. The matriarchal Minangkabau of Sumatra continue to rejoice at the birth of a girl, continue to have women as the head of the households, yet are Muslims. Ideas which seem irreconcilable are routinely incorporated into one, complex life-view. No doubt the anti-Darwinians of 100 years ago would be shocked at the ease with which we reconcile the Genesis version of creation with modern theories of physics.

In ancient Greece, reproductive confusion abounded. We can see this from a trilogy of plays, the *Oresteia*, written by Aeschylus. It concerns the real-life drama of the family of King Agamemnon who commanded the fleet that set off to rescue Helen and destroyed the city of Troy in 1193 B.C. Agamemnon had been forced to sacrifice his daughter, Iphigenia, to appease the gods, confirm his command and ensure a following wind. On his return he is killed by his angry wife and she, in turn, is killed by their son, Orestes. The third play in the trilogy, *Eumenides*, first performed at the theatre of Dionysus in 458 B.C., concerns the trial of Orestes. In it, the god Apollo defends Orestes from the charge of matricide by saying: 'The mother is not the true source of life' but just a 'furrow' where the father thrusts his seed.[15]

As his chief witness, Apollo calls the goddess Athene who, in pre-Hellenic myths, was said to have been born from Lake Trionis but later on was said to have been born from Zeus' head (after he'd swallowed her mother, Metis). In the play, Athene, who stands as the authority of the old matriarchal ways, is made to say: 'I recognise no mother ... I take the male part, strong for the father.'[16] The issue is settled: there is no mother, therefore no charge of matricide. Orestes is acquitted. By means of this drama-propaganda, the message is put across to a confused public: the father is the 'real' parent.

Two years after this play was first performed, Aeschylus died. According to an apocryphal story, Aeschylus met his end when a vulture or eagle dropped a stone (or turtle) on his bald head, thinking it was the large egg of a flightless bird.[17] If the story is true, it was poetic justice because Aeschylus played an important part in the destruction of the idea that women contained the seed – the egg – from which human life proceeded. If the story is just pure myth, it proves that people well understood what Aeschylus had done.

8 An all-encompassing goddess or mere fertility cult?

Some of the strongest evidence we have that women were once thought to be the ultimate source of life comes from the vast number of ancient goddess figurines. These are repeatedly referred to as 'fertility goddesses', as if women went to them only when they wanted a baby, or even, by extension, when they wanted their animals and crops to multiply. However, this is not the way the goddess was, and is, thought of. The Mother Goddess of the ancient past was the source of *everything*, in the same way that today Jahweh, Jehovah or Allah are thought to be the source of all.

In the second century A.D., the Roman writer Apuleius wrote the *Golden Ass*, in which the goddess speaks:

> I am Nature, the universal Mother, mistress of all elements, primordial child of time, sovereign of all things spiritual, queen of the dead, queen also of the immortals, the single manifestation of all gods and goddesses that are. My nod governs the shining heights of Heaven, the wholesome sea breezes, the lamentable silences of the world below. Though I am worshipped in many aspects, known by countless names and propitiated with all manner of different rites, yet the whole round earth venerates me.[1]

The Kagaba of Columbia, South America, still worship the goddess. To them, she is:

> The mother of all our songs, all our seed, bore us all in the beginning – all types of people. She is mother of thunder, streams, trees, the world, and our ancestors the stone people, and all strangers. Mother of our dances, temples, fire, sun, stars and rain. She has left us a token in all our temples in the form of songs and dances.[2]

Of North American Indian spiritual concepts, Paula Gunn Allen tells us that, although 'contemporary Indian tales suggest that the creatures are born from the mating of sky father and earth mother', there are 'older, more secret texts'. Before the masculinist revisions, which Allen says may have pre-dated Christianity or occurred since its influence, the goddess alone was the source of all:

> Thought woman is not a passive personage: her potentiality is dynamic and unimaginably powerful. She brought corn and agriculture, potting, weaving, social systems, religion, ceremony, ritual, building, memory, intuition, and their expressions in language, creativity, dance, human-to-animal relations, and she gave these offerings power and authority and blessed the people with the ability to provide for themselves and their progeny. Thought Woman is not limited to a female role in the total theology of the Keres people. Since she is the supreme Spirit, she is both Mother and Father to all people and to all creatures. She is the only creator of thought, and thought precedes creation.[3]

It has been extremely difficult for sexist academics to grasp the fact that once upon a time (and still) people thought in terms of a female deity who was as all-encompassing as the male deity. One can understand why. We are all products of our time and this is the time of the male deity. Erasing this fundamental concept from our psyche so we can approach the past with a blank sheet, as it were – with no pre-conceived notions – is difficult. Even an atheist academic would, I bet, find himself saying, 'Oh, God, help me', if on a field-trip perhaps, he found a bull elephant charging at him with ears flapping! Could he then say that he approaches the spirituality of ancient peoples with an unbiased opinion? In all other scientific fields of study there are convoluted procedures one must follow to ensure an unbiased approach to the subject material, but in archaeology, which is very much to do with the study of spiritual concepts, no such measures are applied. Yet all human beings are spiritual and so all approach this particular area of study already biased. Archaeologists should be made to perform Goddess rituals and circle dances; they won't be 'converted', but the emotions they experience might just awaken them to their prejudices against the concept of an all-encompassing goddess!

Egyptian goddess
Predynastic terra-cotta figurine with bird head and upraised arms.
(British Museum, London, England)

Not only does calling the ancient goddess a 'fertility figure' do
away with the necessity of thinking in terms of an all-encompassing
deity, it pigeon-holes womanhood into the role of baby-machine.
It says, 'people worshipped the female for her ability to give
birth', and even though this definition is often extended to

include the multiplication of animals and the growth of plants, it is still limiting. It denies the goddess her other aspects: as bringer of death and regeneration (reincarnation), law-maker, prophetess, inventor, healer and warrior. The goddess was the source of wisdom, magic, love, fate, inspiration, change and spirituality – none of which need have anything to do with making babies or making plants grow, but all of which were denied to women over the past two thousand years and more. This is no coincidence. Having denied woman intellectual brilliance and deep spirituality for so long, we find it very difficult to see her in any role other than 'mother' – the flesh and blood, the 'nutritive soul' of Aristotle, the baby-making machine of Luther, or the 'fertility goddess' of archaeology.

This sexism becomes patently clear once you begin to look at large numbers of goddess figurines. From all the talk about a 'fertility goddess', you'd expect, perhaps, to find these images either pregnant, giving birth or holding a baby. But the vast majority are not. It is much more common to find goddess figurines holding serpents, lights, flowers or swords. Elise Baumgartel, who has excavated many early Egyptian sites, writes: 'No figure of a woman with child ... is known to me from predynastic Egypt.'[4] Colin Renfrew writes of the canonical Cycladic figurines: 'A few of the figures are shown as distinctly pregnant, others have a swelling that might suggest pregnancy, but the vast majority, while undoubtedly female, suggest no hint of pregnancy. This observation militates against them representing a fertility goddess.'[5] Precisely.

Far from being the nice, soothing mummy, to whose breasts one can run when in danger, in her association with carrion-eating birds and animals the goddess was often portrayed as danger itself. The goddess was feared in the same way we fear the male gods Jahweh, Jehovah and Allah. And with that fear went respect.

Because most archaeologists can't grasp the fact that the goddess represented birth *and* death (and regeneration), portrayals of her in her dark, negative aspects are called 'a mystery'. Having categorised all female images as having to do with reproduction (and nothing else), their imaginations are clipped. The best they can manage is a side-stepping. So, for example, Baumgartel will redefine the 'mother goddess' as a 'fertility goddess, since she is never represented with a child'. The fact that she isn't represented holding a head of corn either seems to be immaterial. 'Fertility goddess' is the best she can come up with given the reluctance to think in terms of an all-encompassing female deity, and given the other options on offer. Of the Egyptian figurines made without arms (figurines without arms were an extremely common phenomenon in sites all over the world) Baumgartel writes: 'It has been suggested that the female statuettes were concubines for the dead; but they were found with women even more frequently than with men, which seems to dispose of this suggestion. They also have been explained as servant-figures but servants without arms seems to defeat with their own purpose.'[6]

The lady of Pazardžik
Clay seated goddess, 15 cm high, dated at 4500 B.C. Found at Pazardžik, Bulgaria. (Sofia Museum, Bulgaria; cast in Naturhistorisches Museum, Vienna, Austria)

Funny, isn't it, how anything female is assumed to be a 'servant' or a 'concubine' or, as G.H. Luquet thought, an aide-memoire to *men's* 'sensual satisfactions'.[7] Funny too how each time you see the word 'goddess' it is prefixed with the words 'mother' or 'fertility'. Such limited perceptions of the goddess reflect, of course, our limited perceptions of womanhood.

Some people think the female figurines were used in sympathetic magic rituals, particularly as requests to the 'fertility goddess' to make individual women pregnant. However, in ethnographic examples that we know of today, women are more likely to use a model of the desired-for *child*. The first Egyptian figurine holding a child is dated at around 2120 B.C., after the invention of writing and, indeed, a woman has written on it her desire for a child. So why shouldn't earlier figurines also show the baby? Or at least show the woman pregnant or giving birth? And if the prehistoric figurines show the desired-for child grown up, why was the desired gender female?

It's often said that female representations need not necessarily imply goddess worshipping, nor an enhanced respect for the feminine, because today one can find millions of 'female figurines' in the form of the Virgin Mary, yet the context is, as we know, patriarchal. I don't know how this argument makes it into print. The first difference is that the Virgin Mary invariably holds a baby boy, Jesus, and it is her relationship with that boy that gives Mary her religious authority. I know of *no* prehistoric female figurines that hold a baby boy. Invariably, the female figurine is alone or, as in the case of a few Cycladic figurines, for example, she holds a small *female* figurine.

The other difference is that the Virgin Mary figures exist alongside figures of men. Aside from all the images of Jesus on the cross, which can take the form of church or street sculpture, or neck pendants, there are millions of sculptures of male spiritual, political and military leaders. Future archaeologists will have no doubt when digging up our culture that men were pre-eminent. My town, London, is packed with male imagery, the notable exception being the sculpture of Boudicca at Westminster Bridge. But from the chariot she rides, it will be obvious that she came from an earlier time. Indeed, this queen of the Celtic Iceni tribe, who tried to fight off the Roman invasion, came from a long line of fighting women – women who worshipped a Celtic goddess.

For many years archaeology belittled worship of the female by saying any 'goddess' was seen as 'protectress of the hearth' because female images were often found in homesteads – caves or huts or little houses. The implication was that when people started *proper* spiritual worship and built large buildings in which they could gather together as a community and worship, they would be worshipping a male deity. Then in the late 1980s an amazing discovery (to some) was made. At Novali Cori on the Syria–Turkey

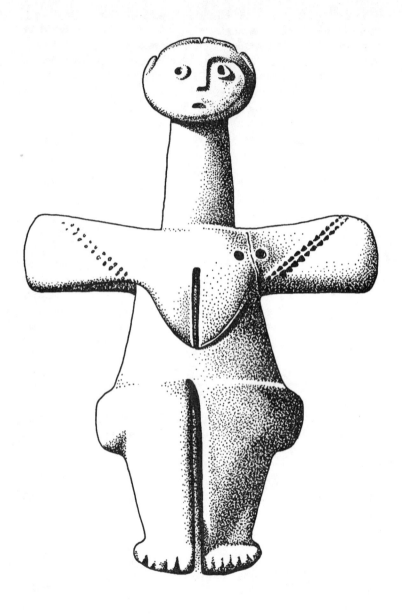

Cyprus goddess
Limestone figurine, 39.5 cm high, dated around 3000 B.C. (J. Paul Getty
Museum, Malibu, California, USA)

border, the earliest known cult centre was found. The people worshipped a pregnant goddess, here associated with a tortoise. The site is 9,000 years old!

Cretan goddess
Clay figurine, dated around 1350 B.C. from Gazi in post palatial Crete. Note the poppy-seed capsules on the head-dress, probably indicating the use of narcotics. (Archaologischen Institut, Berlin, Germany)

The evidence for the ancient worship of a female deity (or deities, if you prefer) is absolutely overwhelming. She was worshipped at cult centres and home shrines. She was fashioned in small, portable images, and sculpted in life-size statues for communal worship outside. She was put in graves with the dead, and in flint-mines with the living. In 1850 B.C., people working deep within the earth at Grimes Grave in Norfolk, England, worshipped a goddess made of chalk, which they placed on a ledge and made offerings to.

The goddess was made of clay, bone, stone, marble, copper and gold. She was portrayed as maiden, mother and crone. She was not simply a 'fertility cult' or the 'protectress of the hearth', she was the mother of all, the source of wisdom, the 'sovereign of all things spiritual' and she 'who weaves us together in a fabric of interconnection'.

It is unlikely that all the images that come to us from the ancient past represent the goddess. Some are undoubtedly priest-esses that served the goddess, while others represent worshippers. I think some of the armless figurines buried with the dead could represent a joint image of goddess and worshipper. They say, 'I am one of yours. I recognise you as my Creator and I am helpless in your presence … take me into your realm.' But whatever, we can't even begin to unravel the enigma of the figurines until we are prepared to make that most difficult leap of the imagination: from an all encompassing male God, to an all-encompassing female Goddess.

9 The phallus

The phallus has its own history. In our culture it will be remembered as the dildo, an instrument to generate sexual pleasure. We didn't invent dildos; they're known the world over. Leading erotic art expert, Philip Rawson, tells us that in China they were made out of ivory or glazed ceramic, which could be filled with warm water. Alternatively, the fibrous stalk of a certain plantain could be soaked in warm water so it swelled to a pleasant bulk and texture. Anthropologist Edgar Gregerson reports that the Chukchi women in Siberia 'are said to use the calf muscle of a reindeer'.[1] Dildos were used in the Middle Ages in Europe, and their use probably goes back a great deal further than that throughout the world.

Today, the phallus as fertility object is extravagantly displayed in the Shinto shrines of Japan. Dozens of huge upstanding members attest the fertilising aspect of man, alongside the Torii arch – emblem of the female aspect. These temples are usually set in beautiful gardens, and couples hoping for a child can spend the night there. These are living places, where sex and religion meet. Not so long ago in Japan, farming communities used to hold fertility dances, then decamp to the paddy fields for the real thing.

The same impetus was behind the raunchy Dionysian rites of ancient Greece and Rome. As in Japan, huge phalluses were carried in processions. In Greece they went to the temple, and in Japan to the fields, where the phalluses were burnt in offering to the sun, that it might return again.

The cult of the phallus must have been most impressive on the Greek island of Delos where, in the third century B.C., there was an avenue of huge stone phalluses atop pillars. They were dedicated to Priapus, son of Dionysus. Apparently, Roman matrons used to sit on smaller versions of Priapus's erect penis to become fruitful.[2]

Going back in time to when man was the stimulator, rather than the generator, the phallus takes on other roles again. According to Elizabeth Gould Davis, 'phallus worship was decreed by women as part of the goddess cult throughout the world'. She explains the ancient Egyptian myth of Isis, once the primary deity:

> When Typhon murdered her consort, Osiris, and cut him up into little bits, Isis went about gathering up the pieces. But nowhere could she find the missing penis. She therefore ordered a wooden lingam to be made, and this she set up in her chief temple at Thebes. It was for this reason that all the goddess temples in Upper and Lower Egypt were adorned with wood or stone phalluses.[3]

To Davis, the abundance of female images in prehistoric times and the comparative dearth of male imagery – except for the phallus – indicates that 'the only thing about a man that was to be valued was his sex organ, made for her pleasure and fulfillment'.[4]

In the story above, Isis could have ordered the incorporation of the phallus into ritual because she knew no phallus = no babies. That doesn't mean, however, the male's contribution was seen as 'seed'. According to Marija Gimbutas, the 'phallic obsession' only became manifest after about 6500 B.C., when ithyphallic gods and phallic-shaped stands and cups begin to make their appearance. She gives another perspective: 'The ubiquity of phallic symbols connotes the glorification of the spontaneous life powers. Phallicism certainly had no obscene allusion; in the context of religious ritual it was a form of catharsis, not of symbolic pro-creation. There is no evidence that in Neolithic times mankind understood biological conception.'[5]

During the goddess-worshipping days, phalluses were placed in front of the images of the goddess, such as at Grimes Graves in England in 2850 B.C. (Only later did the phallus itself become the object of veneration.) Phalluses were offered to the Cretan goddess Artemis, for example, who, aside from protecting women giving birth, was the protectress of animals. She struck a powerful figure with her bows and arrows, flanked by her hounds and stags.[6]

Probably the oldest philosophical tradition still in existence is Tantra. Shrii Shrii A'nanda Mu'rti describes it thus: 'Tantra is not a religion, but a way of life, a system of sa'dhana'. The fundamental goal of this sa'dhana' is to awaken the dormant jiivashakti (unit force), known as kulakan'd'alinii, and, after elevating it stage by stage, to merge it in Brahmabha'va (Cosmic Consciousness).'[7]

Cosmic consciousness was what people talked about thousands of years ago, before the Aryans thundered into India. Tantra was strong in Bengal and elsewhere, but it may also have had roots in the Indus Valley Civilisation which, from 5000 to 1900 B.C., stretched in a band about 800 miles wide and 2000 miles long across present-day Pakistan, and also along the Pakistan and north-western Indian shores. One of the ways to reach Cosmic Consciousness is to have Tantric sex, and the people of the Indus may have realised this, at least it seems so from the things they left behind.

These people were great traders and were known in Sumeria and even as far, it's now thought, as Ireland. They had a great civilisation. In one of their cities, Mohenjo-daro, each house had its own bathroom and toilet, which were connected to the public drainage system. Built on a grid system, this city of 40,000 people was obviously highly organised. Of its spiritual life, Arthur Cotterell writes:

> The lack of imposing temples supposes worship at family altars, and a number of small images in clay, stone and bronze have been recovered at Mohenjo-daro. Conspicuous among them are representations of a mother goddess, the Indus equivalent of Inanna or Isis. As the universal mother, she bestowed fertility on plants, animals and men. Her cult is especially represented by carved sexual symbols: upright phallic stones, denoting her consort, and circular stones with a hollow centre, representing her own teeming womb. They may be the primitive prototypes of the Hindu symbols of the *lingam* and the *yoni*, which are common today in the temples of Shiva and his goddess Devi.[8]

These 'primitive prototypes', as Cotterell calls them, didn't have the same meaning in 5000 B.C. as they did in later, male-seed, times. Indeed, as time passed and semen-as-seed came to be all-important, Shiva's Lingam grew so enormous that it reached from the upper stratosphere to the core of the earth, so the stories tell us.

According to Philip Rawson: 'There is a considerable amount of evidence that kings in early India, as well as in Indianized regions of Southeast Asia, owed their kingship to marriage or sexual union with female "owners" of the terrain.'[9] I think women were caretakers of the land, acting on behalf of the goddess. According to Shrii Shrii A'nanda Mu'rti, a remnant of the Tantric matrilineal social system which existed in Bengal before the Aryan invasions can still be seen in the Dravid-Keralite and Mongolo-Khashian societies.[10] He also says: 'The speciality of non-Aryan ceremonies is that the women's role is predominant.'[11]

What went through the minds of people who, up to 7,000 years ago, venerated the penis and vagina? Perhaps they knew that through spirituality, induced by yoga-sex, there was Cosmic Bliss. Perhaps they thought, 'we are all One, in the Great Goddess'. Gender could have been seen as illusion, just a manifestation of the deeper truth. Or perhaps the lingam and yoni simply represented the understanding that life proceeds from intercourse, the locking of opposites.

The archaeologist Marija Gimbutas refers to an 'Old European' culture, which stretched from France to Russia, as far south as the Aegean and Mediterranean seas. It dates from at least 6500 B.C. and was brought to an end on the mainland when the Kurgans invaded between 4300 and 2800 B.C., but continued along the Mediterranean coast and on the islands until 1500 B.C. Of it she writes: 'The Old European phallus is far from being the obscene symbol of our days. Rather, it is close to what is still found in India, the *lingam*, a sacred cosmic pillar inherited from the Indus Valley civilisation.'[12]

She is not referring to the later vast rod of Shiva, which had the other gods and goddesses cowering in awe, but the original lingam, which she seems to interpret as the organ of man the stimulator: 'Phallic cult articles ... do not represent a male god but rather a vivifying and fructifying force of nature appearing as an aspect of life column symbolism.'[13]

The word 'vivify' means 'to give life to; to endue with life; to animate; to quicken'. The first two meanings here give us a clear picture of the crucial nature of the male role – he gives life. Yet the second two definitions clarify the nature of that giving – he animates. One animates something that is already present – the seed which grows into the plant one can see displayed across the abdomens of female figurines. We have an example of this from a site in Pakistan, Sheri Khan Tarakai, which dates from between 4500 and 3000 B.C.

The most ancient monumental architecture still existing can be found on the Mediterranean islands of Malta and nearby Gozo. Aside from the Hypogeum at Hal Saflieni – a series of egg- or womb-shaped underground chambers, dated 4000–2500 B.C. – there are 23 temples built on the shape of the female body. Veronica Veen says that: 'The temple as a big womb in which even a whole community could seek renewal, is able to make the social structure even more coherent. A dark cave-like womb, which you enter by the trilithon, the vagina of the goddess, and out of which you will come reborn.'[14]

J. Bezzina discusses the phallus in relation to these temples:

> It was later on in the neolithic period that man became aware of the male's part in procreation. The Phallic symbol was introduced in the last part of the temple period, though it never gained a place in the 'Holy of Holies' where the goddess was dominant. It was found either outside the temple in a deliberately built-up niche, in the exterior wall (Haġar Qim) or in the interior, facing the Holy of Holies (Tarxien) or in one of the apses near the sacred place. In all cases it is facing a downward-tapering slab or the Pubic Triangle, thought to represent the female. From these symbols it is understood that fertility was very much involved.[15]

During the final episode of goddess temple culture, we see the occurrence of little stone phalli, sometimes arranged as tiny altars. This practice may have been introduced by a sub-culture, or it may indicate a loss of confidence in the goddess as a result of bad growing conditions and malnutrition. As if to try and re-assert her position, between 2700 and 2500 B.C., followers of the goddess carved out a huge figure 2.5 m high and enormously wide. It is the largest Neolithic goddess figure known to exist. Smaller figures had been carved from earliest times in limestone, and smaller ones still, moulded in clay. An interesting feature of Malta's goddess past is that many figurines have a socket between the shoulders, to which a variety of heads could be attached. Bezzina suggests the head was carved separately from the body so 'it could be changed for different ceremonies, rites or even seasons'.[16]

Going back in time again, the phallus all but disappears. Only two examples are known from Paleolithic times (one of which doesn't look like a penis at all but male archaeologists like to think it is), compared to over two hundred examples of female figurines from the same period. The penile shape still exists, but it is attached to the shape of a female body. A classic example of this tradition comes from Savagnano in northern Italy. It's a typical goddess figurine from the shoulders to the ankles – complete with large pendulous breasts, protruding tummy and ample buttocks

– but the head is in the shape of a short, pointed penis, complete with shaft. It was carved in serpentine marble in 20,000 B.C. and stands 23 cm high (illustrated on p. 164). A piece dated 24,000 B.C. from Dolni Vestonica in Czechoslovakia, is simply a baton with large breasts which, because of the proportions, looks like a female body with a phallus for a head. I can't see that the figurine recovered at Savagnano is essentially different from some of the Cycladic figurines dated 3000 B.C., which have typical female bodies but a shaft instead of a head. Another, from Charente in France, dated between 15,000 and 13,000 B.C., is a long pointed penile shaft extending from a vulva, set in the woman's upper thighs (illustrated on p. 165).

What we're talking about here are very powerful and very sexy images. They have a raunchy quality which is slightly disturbing. I think that in some cases at least, ritual defloration was involved, and I discuss that in a later chapter, but these objects could convey a concept little discussed in archaeological circles, that Paleolithic and Neolithic people might have appreciated the empowering nature of sex. But whatever idea these objects were designed to convey, they existed in a female-centred culture. Most female figurines have heads, not phalli.

10 Woman the boss

It is a myth that men have always been boss. In what may be the oldest book in existence, the 4,500-year-old *Maxims of Ptah Hotep*, men were advised:

> If you're wise, stay home, love your wife and don't argue with her. Feed her, adorn her, massage her. Fulfil all her desires and pay attention to what occupies her mind. For this is the only way to persuade her to stay with you. If you oppose her, it will be your downfall.[1]

Whatever it might sound like, Ptah Hotep was no wimp. He was Vizier in Egypt, and carried the burden of government from around 2420 to 2400 B.C., during the Fifth Dynasty, when Egypt was already an ancient civilisation. His sayings were handed down as literature and no doubt further copies of his work would have been in the 'classics' section of that fabulous repository of ancient wisdom, the great library at Alexandria, before it was burnt to the ground by a horde of Christian fanatics in A.D. 391. I wonder what other interesting gems went up in smoke on that tempestuous day.

I don't know a man living who would sign this, the groom's pledge from a first century B.C. marriage contract, discovered by the Greek historian Diodorus Siculus in Egypt:

> I bow before your rights as wife. From this day on, I shall never oppose your claims with a single word. I recognise you before all others as my wife, though I do not have the right to say you must be mine, and only I am your husband and mate. You alone have the right of departure ... I cannot oppose your wish wherever you desire to go. I give you ... (a list of the bridegroom's possessions).[2]

In Egypt, according to Diodorus, reverence of the goddess Isis accounted for the fact that: 'It was ordained that the queen should have greater power and honour than the king and that among private persons the wife should enjoy authority over the husband, husbands agreeing in the marriage contract that they will be obedient in all things to their wives.'[3]

In the adjacent country of Libya, Diodorus found that:

> All authority was vested in the woman, who discharged every kind of public duty. The men looked after domestic affairs just as the women do among ourselves and did as they were told by their wives. They were not allowed to undertake war service or to exercise any functions of government, or to fill any public office, such as might have given them more spirit to set themselves up against the women. The children were handed over immediately after birth to the men, who reared them on milk and other foods suitable to their age.[4]

In Babylon, according to nineteenth-century scholar W. Boscawen, the sign for mother was 'goddess of the house'.

> The freedom granted to the women in Babylonia allowed them to hold and manage their own estates and this was especially the case with priestesses of the temple, who traded extensively ... One of the most interesting and characteristic features of this early civilization of the Babylonians was the high position of women. The mother here is always represented by a sign which means 'goddess of the house'. Any sin against the mother, any repudiation against the mother was punished by banishment from the community. These are the facts which are evidently indicative of a people who at one time held the law of matriarchal descent.[5]

The suffragettes had a lot more to fight for in the early twentieth century than women in ancient Egypt. Barbara Lesco describes the evidence from 1567 to 1085 B.C., known as 'the New Kingdom':

> Numerous texts have survived from this period, including court documents and private letters revealing that women had their own independent legal identity on a par with men and that they could inherit or purchase property and dispose of it without a male co-signatory or legal guardian. Indeed, women were the heads of households, testified in court, witnessed documents, acted as executors of their family estates and assumed the obligations of a citizen *vis-à-vis* the State. Numerous records show this was true of free

women in general, not just those of the gentry. On a personal level, it is clear that women enjoyed freedom of movement and association, that they could marry and divorce at will, that they engaged in commerce and that they were able to exercise authority over others in the work-place or temple.[6]

Ancient women didn't stay home, as they were later ordered to do. According to 'the father of history', Herodotus, who travelled around in the fifth century B.C., in Egypt: 'Among them the women attend markets and traffic, but the men stay at home and weave.'[7] Even as late as the first century B.C., Diodorus found, on the west coast of Turkey, a land where men carried out the domestic duties, including the spinning of wool, while 'women held the supreme power and royal authority'. These people worshipped 'The Mother of All Deities'.[8]

Artefacts recovered from early Greek sites show women 'driving chariots, leading hunts, occupying the best seats at the theatre, presiding in the halls of justice, and receiving homage from men'.[9] Charles Seltman tells us that the Pelasgians of pre-Mycenaean Greece were 'dominated by the female principle; men were but the servers of women in the chase, in the fields, in love, and in war'.[10]

Clearly, women were not always under the control of men and, as we shall see, in ancient times and in many places women had more freedom than they do now in terms of spiritual and public authority, sexuality and intellectual creativity.

Spiritual authority and queenship

In 2680 B.C., women had more access to spiritual authority than they do now, in Egypt at least. According to Barbara Lesco, during the Old Kingdom, 2680 to 2160 B.C.:

> It would seem that few restrictions were placed on women of ability and high social status ... It is interesting to note that religious positions were not limited to noble-women, for we have found priestesses of major goddesses who bear humble titles such as tenant farmer. Many administrative, honorific and priestly titles for women have been recovered from Old Kingdom monuments.[11]

Lesco explains Egyptian sculpture:

> The fact that the king's family were often depicted at the
> level of the legs of his colossal figure should not blind us
> to the importance of the great royal wife in Egyptian
> history. The queen was often of purer royal blood than her
> husband, whose claim to the throne she legitimised.[12]

Kingship started with queenship. The latter developed from the
position of high priestess – she who mediated with the goddess,
and administered the land on her behalf. As administrators of the
goddess temples, women controlled vast estates and numbers of
personnel. In the Sumerian city of Lagash, for example, in the year
2300 B.C., the temple of the goddess Bau employed up to 1,200
permanent workers. The powerful temple connection ensured that
women were employed as judges and magistrates in Nimrud,
northern Mesopotamia, and no doubt elsewhere, as late as the
eighth century B.C.

In her ground-breaking work *When God was a Woman* Merlin
Stone has found plenty of evidence confirming the prime
importance of women during early civilisation. It was once
thought that Sumerian kings appointed their wives to the position
of chief priestess but we now know the priestess decided who was
to become king. According to Sumerian documents, he attained
his position once he had 'proved himself' in her bed; that is, the
bed of the goddess, represented by the priestess. Thus, mystically
identified with Damuzi, the kings of Sumer were known as the
'beloved husbands' of Inanna.

Sumeriologist Dr S.N. Kramer says the priestess of Inanna was
the 'dominant partner' in her relationship with the king. According
to Dr Sidney Smith, 'no king acted according to his own judgement
alone'. Before he did a thing, the king had to consult the priestess
who, after some oracular divination and prophecy, would give
the instructions.

Of Mesopotamia, Professor Henri Frankfort says, 'the goddess
is supreme because the source of all life is seen as female. Hence
the god too descends from her and is called her son, though he
is also her husband.'[13] To the east of Sumeria there was a nation
called Elam, of which Dr Hinz wrote: 'Pride of place in this world
was taken by a goddess – and this is typical of Elam ... She was
clearly the "great mother of the gods" to the Elamites. The very
fact that precedence was given to a goddess, who stood above and

apart from the other Elamite gods, indicates a matriarchal approach in the devotees of this religion.'[14]

Around 6500 B.C. in Çatalhöyük, Turkey, 'the supreme deity in all the temples was a goddess', 'women were the heads of households' and men were 'utterly subject to their mothers', wrote James Mellaart.[15] Of the figurines he excavated in Hacilar, dated 1,000 years later, he says: 'The statuettes portray the goddess and the male occurs only in a subsidiary role as child or paramour.' Arthur Evans wrote: 'Throughout a large part of Anatolia, again we recognise the cult of the same great mother with her male-satellite husband, lover or child, as the case may be.' He also said the goddess ruled supreme in Crete.

According to H.J. Rose, early Greek myths: 'represent the goddess, not as married but as forming more or less temporary unions with someone much inferior to herself, a proceeding quite characteristic of Oriental goddesses who are essentially mothers but not wives and besides whom their lovers sink into comparative insignificance'.[16]

During the earliest days of kingship, the kings were dispensed with, literally, once a year. The high priestess, representing the goddess, chose for herself a handsome young man who, for a year only, would enjoy all the luxuries of life, a certain amount of power, and the sexual favours of the priestess, through whom he would gain access to the divine. He would then be required to take part in the Sacred Marriage – the *hieros gamos*. Rosalind Miles gives an evocative description of one such event:

> The goddess Anaitis of Nineveh annually demanded the most beautiful boy as her lover/victim: beautiful with paint, decked with gold ornaments, clothed in red and armed with the double axe of the goddess, he would spend one last day and night in orgiastic sex with the priestess under a purple canopy in full view of the people, then he was laid on a bed of spices, incense and precious woods, covered with a cloth of gold and set on fire. 'The Mother has taken him back to her,' the worshippers chanted.[17]

Versions of the Sacred Marriage existed in Sumer, Babylon, Egypt, Carthage, Syria, Palestine, Cyprus and throughout the Aegean. Its symbolism has been variously interpreted. Of Crete, Stylianos Alexiou suggests: 'The sacred marriage, the union of the goddess and the god (who usually dies shortly after his wedding) symbolises the fertility of the earth.'[18] One can see how this idea might have come about. Vegetation, and indeed all life, goes through a fallow period. After the harvest or the autumnal fall, life seems dead. Hidden deep within the earth or plant, the seeds of life are waiting for the revival. However, before a revival there has to be a proceeding death. That's where the 'annual king' came in (or went out!); he was sacrificed for the good of the whole community.

There may have been another reason for the disposal of the king. Merlin Stone points out that although the idea of a fertility rite may have been the reasoning behind the Sacred Marriage in later times, the discovery and decipherment of our earliest written records allow us to see another side to the subject. She refers to the 2000 B.C. Sumerian legend of the goddess Inanna, 'probably a written record of even earlier myths and religious ideas':

> In this legend the sacrifice of the consort occurred when he was no longer willing to defer to the wishes, commands and power of the Goddess. This most ancient account perhaps reveals the earliest origins and reasons for the death of the male consort. Later ideas of fertility or expiation of sins may have eventually been embroidered about the custom to ensure or explain its continuation.[19]

Different cultures disposed of their annual kings in different ways. Irish priestesses apparently decapitated them, the sacrificial blood being caught in beautifully decorated silver cauldrons. Legends tell of the annual kings being struck by a thunderbolt immediately after the vital union, or coming to a variety of 'accidental' deaths, such as being killed by a wild boar.

In later times, as the year became 'the longer year' (and eventually hereditary kingship), substitutes were offered: other human beings, animals or effigies like the 'mannikins' the Vestal Virgins threw into the Tiber to 'drown'. There was also the practice of ritual assault, which in later times replaced the actual death of the king. In ancient Babylon, according to L. Frobenius, the system had been weakened 'in as much as the king at the New Year Festival in the temple was only stripped of his garments, humiliated and struck, while in the marketplace a substitute, who had been ceremonially installed in all glory, was delivered to death by the noose'.[20] Merlin Stone elaborates, and asks the pertinent question:

> Various accounts of the ceremonies that took place during Babylonian periods tell of the king going to the temple to

be struck in the face, his clothing and royal insignia temporarily removed. Other texts tell us that his hair was shorn, his girdle removed and in this state he was thrown into the river. When he emerged he was made to walk about in sackcloth for several days as a symbol of mourning ... There are hints of expiation of sins and atonement in these rituals – the king is being punished. But for what? It seems that eventually the chastisement came to be for the sins of the people, but did this not originate from his earlier punishment for refusing to defer to the priestess-queen? The fact that good fortune was predicted if tears came to his eyes when he was struck perhaps reveals these origins. According to the Babylonian tablets, 'If the king does not weep when struck, the omen is bad for the year.'[21]

What we may be seeing in the evolution of kingship is the change in reproductive theory. Men were required for reproduction, that much was known, and a king was appointed. But his role was minor and transitory and, as if to illustrate the point, he was annually dispensed with. Life centred round women – they were perpetual, not he. When he challenged his lowly position, the king was punished. It was a message to all men: she's in charge. Eventually, of course, men came to see themselves as more important and eventually took control. They appointed their wives as high priestesses and there seemed to be continuity. In fact, we'd entered a new era.

Sexual freedom

Men who are concerned about ensuring paternity impose certain controls on women. They don't let them sleep around. Here again, things were different in the ancient past. It was quite common, for example, for women to have sex with strangers within the confines of the temple walls. We have evidence that this happened throughout the Mediterranean, in North Africa, Sicily, Cyprus, Greece, Palestine, Lebanon and Turkey.

Even as late as the first century A.D., according to Lucian, women had sex with strangers at the temple of Aphrodite in Corinth, on the feast day of Adonis. According to Herodotus, in ancient Babylon women only had sex with a stranger once – when losing their virginity. Other women lived within the temple complex and were known as *qadishtu*, which outraged Victorian scholars translated as 'prostitutes'. It actually means 'sanctified women' or 'holy women'. The *qadishtu* didn't need cash. According to Merlin Stone: 'Sumerian and Babylonian documents reveal that these women, through their affiliations with the temple complex, owned land and other properties and engaged in extensive business activities. Various accounts report that they were often from wealthy families, well accepted in the society.'[22]

'Official' encouragement of promiscuous sex does nothing to ensure paternity – and perhaps that was the point. If paternity can't be proved, matrilineal descent patterns are more likely to continue. Or perhaps there was no ulterior motive on the part of the priestesses in charge. Perhaps they genuinely thought a woman's fertility stood a better chance of springing into action if it was stimulated by more than one man.

Scholars are not agreed on whether the festivals associated with the 'Sacred Marriage', including as they did public displays of intercourse between priestess and 'king', encouraged sexuality among the people watching. We know that in ancient Greece there were month-long orgies associated with the worship of Dionysus, the fertility god, and that peoples elsewhere in the world, even today, indulge their sexual fancy on festive occasions. So it's not outside the realms of possibility that promiscuous sex was given official approbation in earlier times.

It's a commonly held belief that because woman is the physically penetrated partner there is a basic psychological need for female surrender and male dominance. I doubt the Sumerians saw it that way. With her stream of young consorts, 'lap of honey', 'boat of heaven' and dishevelled hair, their goddess Inanna was in charge. (As she was in war, with her 'rivers of blood'.) When Inanna gave the Sumerians the secrets of the sacred sexual customs, they did not apparently include the missionary position. Indeed, during prehistory this position is rare in representational art. Intercourse is shown in the rear-entry position, side-by-side, woman-on-top or the 'oceanic' position, where the couple face each other, squatting on the floor. This is the preferred position of the Trobrianders who told Malinowski the missionary position was ineffective because 'the man overlies heavily the woman; he presses her heavily downward, she cannot respond'. She has no *'ibilamapu'* – no control. Male sexual dominance is a cultural peculiarity, not a way of life. And in orgasm, men surrender too; like women, to greater or lesser degree.

Even as late as the fifth century B.C., Herodotus found pockets of promiscuity, social behaviour that would certainly not be allowed in the same places today. In Libya, for example, he found the Gindanes women displaying their leather ankle bands, each representing a different sexual partner. He says 'she who has the most is the most esteemed, as being loved by the greatest number of men'.[23] The Nasamonian bride, also of Libya, had to have sex with all the male wedding guests, each of whom would give her a present. But there was another reason to have promiscuous sex, as the Agathysi of Scythia told Herodotus: 'They have promiscuous intercourse with women, to the end that they may be brethren one of another, and being all of one family, may not entertain hatred toward each other.'[24]

Now there's a novel thought!

Woman the inventor

The 'modern' notion that women lacked reproductive creativity led, inevitably perhaps, to the idea that they were intellectually uncreative too. Women were designated men's helpmates and consigned to a life of domesticity and childcare, while men were allowed the time and space to do other things – like writing symphonies or inventing the automobile. This whole scenario has been projected backwards in time so we are led to believe that men invented everything of importance since time began. This is a chimera, a fantasy, because before women were locked in and shut up, they had control of their environment and their minds were free to create – and create they did. Indeed, law, government and organised religion developed under women's influence, and metallurgy, architecture, textiles, pottery, boats and the wheel were invented long before God told Adam, 'Thou art the head, not she'.

Consider the irony of the fact that 100 years ago in Britain women were denied access to the higher education facilities in which they could learn Pythagorean theory while, over 2,500 years ago Pythagoras himself was taught by women – Aristolea and Themistoclea – and was married to a leading mathematician, Theano. Going back even further in time, if we take away our sexist bias, it seems as if women played a much more important role in all aspects of intellectual development than might have been previously thought.

In *Women in Prehistory*, archaeologist Margaret Ehrenberg writes:

> The discovery of farming techniques has usually been assumed to have been made by men, but it is in fact very much more likely to have been made by women. On the basis of anthropological evidence for societies still living traditional foraging lifestyles and those living by simple, non-mechanised farming, taken in conjunction with direct archaeological evidence, it seems probable that it was women who made the first observations of plant behaviour, and worked out, presumably by long trial and error, how to grow and tend crops.[25]

Interestingly, the *Popul Vuh* of the South American Mayan civilisation, which flourished between A.D. 300 and 1450, also attributes the invention of agriculture to women.

The domestication of animals could also have been a female initiative. Although popular culture might depict the tough, strong hunter dominating an animal into submission, the easier route to animal domestication involves capturing an infant, perhaps after killing the mother, and rearing that infant back at the homestead. In Papua New Guinea, women suckle wild piglets at their breasts and thus the animal is integrated into the household.

The female-centred 'Old European' culture, which dated from 7000 to 3500 B.C., and to 1500 B.C. in some places, was artistic and peaceful. Of it, Marija Gimbutas writes:

> Its people did not produce lethal weapons or build forts in inaccessible places, as their successors did, even when they were acquainted with metallurgy. Instead, they built magnificent tomb-shrines and temples, comfortable houses in moderately-sized villages, and created superb pottery and sculptures. This was a long-lasting period of remarkable creativity and stability, an age free of strife. Their culture was a culture of art.[26]

As far as we currently know, the person who invented writing was an accountant, and a woman. Probably, she was one of the *naditu* priestesses who, around 3200 B.C., kept records at the Temple of 'The Queen of Heaven', Inanna, in the Sumerian city of Erech. From here, we have our first cuneiform tablets, which record land deals and rentals. Because the high priestess was thought to be the goddess incarnate, she and the temple were caretakers of the goddess' land, flocks and much else besides. Professor Sidney Smith tells us that in Sumeria: 'the temple directed every essential activity, not only matters that might be considered religious business, but the urban activities of the craftsmen, the traders and the rural employment of farmers, shepherds, poultry keepers, fishermen and fruit gardeners'.[27]

The temple at Erech (Uruk, or modern Warka) stood in the centre of a complex of buildings from where all this activity was directed. Someone had to keep track of it all and, as 'necessity is the mother of invention', cuneiform tablets were invented. The Sumerians said that the goddess Nibada invented clay tablets and writing, and she was known as 'the learned one of the holy chambers, She who teaches the decrees, the great scribe of heaven'.[28] Nibada was a goddess of Erech so it seems likely that she 'worked' through one of her priestesses. Perhaps the high priestess, the goddess incarnate, relayed Nibada's wisdom while in a trance or under the influence of hallucinogenics or, more likely, Nibada 'worked' through a particularly inventive priestess whose job it was to keep track of the vast temple estates.

All over the ancient civilised world, people attributed the invention of writing to a female: Queen Isis was said to have given the alphabet to the Egyptians; the Hindu goddess of knowledge, Sarasvati, was thought to have invented the original alphabet; while Kali was said to have invented Sanskrit. Legend has it that Medusa gave the alphabet to Hercules, and that Carmenta created Latin from the Greek. The ancient Celts worshipped Brigit as the patron deity of language while the goddess Cerriwed was the source of intelligence and knowledge. Although these ancient stories attribute the crucial communication skills to goddesses, like Nibada in Sumeria they may reflect the involvement of real live women.

In time, however, women were to be denied access to positions in which they could exercise their intellectual capabilities. One of the last brilliant pagan women was Hypatia, who held the chair of Platonic philosophy at the famous University at Alexandria, Egypt. She also taught geometry, astronomy and algebra, and is said to have invented the astrolabe – an instrument for measuring altitudes and solving other problems in astronomy; the planisphere – another astological device; the hydroscope – or water clock; and an aerometer – an instrument for measuring the weight or density of air and gases. In A.D. 415 she was dragged from her chariot by a Christian mob incited by the zealous Bishop of Alexandria, Cyril, stripped naked, and tortured to death by 'slicing her flesh from her bones with shells and sharpened flints'.[29] Women of independent minds and means were going out of style.

Woman the boss

People say to me, 'Surely, men took control, they always do.' Well no, they don't, not in matriarchal societies. You can see this today among the Minangkabau people of Bukittinggi in Sumatra, the westernmost island of Indonesia. Here, the birth of a girl is a 'blessed event' and grandma is boss, as Richard Mahler reports:

> The eldest living female in a family is considered the matriarch and has the most power in the household, which can number as many as 70 people, all descended from one ancestral mother and living under the same roof. She is deferred to in all matters of family politics. Property, though worked and used collectively, is passed on through the female line of the family, inherited by the daughters in the family. All progeny from a marriage are regarded as part of the mother's family group and the father has little if any say in family matters.

And the matriarch has political power too:

> In the community at large, the eldest women in each family get together in large meeting houses to make decisions on a consensus basis. I'm not sure of the exact nature of the decisions they convene to discuss, although I assume they have to do with projects affecting all members of a village, such as water projects, schools and so on.[30]

Quite often, behind a male 'front', there was a female power. The Ashanti warrior kings of Ghana, who repelled invaders from their gold-rich lands from the fourteenth century until 1896, were the active face of power. The spiritual base of that power lay with the king's mother – the Queen Mother. At least it was until King Osai Kojo in the 1770s consolidated his new unity of kingdoms and started encouraging the appointment of men of ability, rather than men who wielded power just because they were their mother's sons.

When the invaders started bothering the North American Indians, it was male chiefs they met. These could have been the 'red chiefs' behind which were the 'white chiefs':

> There is an old tradition among numerous tribes of a two-sided, complementary social structure. In the American Southeast this tradition was worked out in terms of the red chief and the white chief, positions held by women and by men and corresponding to internal affairs and external affairs. They were both spiritual and ritualistic, but the white chief or internal chief functioned in harmony-effective ways. This chief maintained peace and harmony among the people of the band, village, or tribe and administered domestic affairs. The red chief, also known as the war chief, presided over relations with other tribes and officiated over events that took people away from the village.[31]

Who would have guessed from looking at cowboy movies that the Iroquois, for example, were a 'mother-centred, mother-right people'? According to Paula Gunn Allen, their 'political organisation was based on the central authority of the Matrons, the Mothers of the Longhouses (clans)'. Interestingly, if a woman took her husband's name after marriage, she became ineligible for appointment to the Matron's Council.

As we can see from American Indian history, from the Ashanti and the Minangkabau, there are different ways in which women can hold power. In western society, long after most of women's important duties and privileges had been stripped from them, a strong woman could still become a prophetess. The 'Pythian' at the oracle of Delphi was a woman and long after the kings and other male leaders had the world under their iron grip, they would still consult with her and other prophetesses before making any crucial decisions. Earlier, she would have had veto over their activities but, either way, for millennia, women as oracular prophets had a profound influence on the affairs of state.

The oldest evidence we have for communial human activity, aside from hunting, comes from Moldavia, in north-eastern Romania, and is dated at 4700 B.C. It's an extraordinary collection of 21 little figurines, arranged within a vase, and is known as 'the assembly of snake goddesses'. The collection consists of three highly decorated seated female figurines, some less decorated female figurines and some smaller undecorated figurines, plus two small objects which might indicate that this scene represents a ritual. It could depict young women being initiated into womanhood, or it could simply show a group of women talking, singing, chanting or whatever. In a sense, it doesn't matter what they're doing. Whatever it is, they're doing it together and, presumably, for the common good. There are several other prehistoric collections like this and the interesting thing is that, as far as it is possible to see from these very ancient objects, there isn't a male among them. Women are depicted doing things together, not men. It is possible, I suppose, that men before 3000 B.C. were the bosses and organised communal activities (aside from hunting). However, there isn't a scrap of evidence to support that idea.

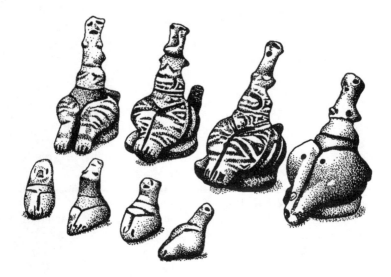

The assembly of snake goddesses
A collection of 21 figurines, ranging from 6–12 cm high, found stored in a large vessel at Poduri-Dealul Ghindaru, Moldavia, northeastern Romania. Dated between 4800–4600 B.C. (Piatra Neamţ Museum, Moldavia, Romania)

11 Portrait of a takeover: Yahweh

In the beginning, the Hebrew people worshipped a goddess. Biblical scholar Raphael Patai says that there can be 'no doubt that to the very end of the Hebrew monarchy the worship of the Old Canaanite gods was an integral part of the religion of the Hebrews ... the worship of the goddess played a much more important role in this popular religion than that of the gods'.[1]

The prophets Jeremiah, Ezra, Nehemiah and Hosea spent most of their time abhorring this 'abomination'. In the Old Testament (Jeremiah 44), we hear how Jeremiah went to Egypt to try and convince the Jews living there to stop their evil-doing and threatened them, on behalf of the God Yahweh, with punishment 'by the sword, by the famine and by the pestilence'. The crowd at Pathros were not impressed. They replied:

> ... we will not hearken unto thee. But we will certainly do whatsoever thing goeth forth out of our own mouth, to burn incense unto the queen of heaven, and to pour out drink offerings unto her, as we have done, we, and our fathers, our kings, and our princes, in the cities of Judah, and in the streets of Jerusalem: for *then we had* plenty of victuals, and were well, and saw no evil. But since we left off to burn incense to the queen of heaven, and to pour our drink offerings unto her, we have wanted all *things,* and have been consumed by the sword and by the famine.(Jeremiah 44: 16–18)

Yahweh reacted with vengeance: 'I will watch over them for evil, and not for good: and all the men of Judah that *are* in the land of Egypt shall be consumed by the sword and by the famine, until there be an end of them' (44: 27).

'The Queen of Heaven' was also known as the 'mother of all deities', 'celestial ruler' and the 'mistress of kingship'. She was worshipped as Astarte, Ashtoreth, Attoret, Anath, Elat and Baalat. So-called 'Astarte plaques' are the most common religious objects recovered from the biblical lands. The idea for them was borrowed from Mesopotamia, where they date from at least 3200 B.C., and they were used at least until the seventh century B.C. in Tell Beit Mersim, the biblical town of Devir, south-west of Jerusalem. Here is a description of them:

> These are pottery plaques, generally oval in shape, on which were impressed (from a pottery or metal mould) a figure of the nude goddess Asherah, *en face* with her arms upraised, grasping lily stalks or serpents, or both, in her hands. The goddess's head is adorned with two long spiral ringlets identical with the Egyptian Hathor ringlets.[2]

I wonder if, as part of the drive to 'disappear' the opposition, these ringlets were adopted by Jewish men and boys in symbolic acquisition of the goddess' creative powers. If so, Hassidic men still wear such ringlets today.

The 'disappearance' of the goddess took many forms. She was referred to in the masculine or repeatedly linked to her male consort, Baal, as if he were the main force in the 'alternative' religion. Such belittling was so effective that, over time, we have come to believe that 'gods' were the problem and 'Baalism' the religion Yahweh was fighting against. It's clear from the few descriptions of actual ritual in the temples, that the goddess religion was alive and well in biblical times. But, if everyone did what they were told in Deuteronomy 12: 2–3, it wouldn't be for long:

> Ye shall utterly destroy all the places, wherein the nations which ye shall possess served their gods, upon the high mountains, and upon the hills, and under every green tree: And ye shall overthrow their altars, and break their pillars, and burn their groves with fire; and ye shall hew down the graven images of their gods and destroy the names of them out of that place.

So hateful was the worship of the goddess that it warranted murdering your own child:

> If thy brother, the son of thy mother, or thy son, or thy daughter, or the wife of thy bosom, or thy friend, which *is* as thine own soul, entice thee secretly, saying, Let us go and serve other gods ... thine hand shall be first upon him

> to put him to death, and afterwards the hand of all the
> people. (Deuteronomy 13: 6,9)

Although killing a wife might be called justifiable homicide under such circumstances, you notice there is no mention of killing a husband. The divine directive to kill was given to men.

What could possibly justify such murderous behaviour? According to biblical scholar I. Epstein, the problem was sex:

> It is important to understand that the vehement opposition
> to idolatry which distinguishes the legislation of the Bible
> and later of the Talmud was not merely antagonism of one
> theological system to another. Fundamentally it was a
> conflict of ethical standards. Heathen people practised
> abominations against which the scriptures earnestly warned
> Israel. Idolatry was identified with immoral conduct, an iden-
> tification which was too often verified by experience.[3]

The 'immoral conduct' was actually built into the goddess religion in the form of the 'sacred sexual customs', and matrilineal descent was intrinsic to it. Children belonged to their mothers and because of female promiscuity, paternity was none too sure. All this was an 'abomination' indeed to the Levite priests who were trying to pull the Hebrew tribes into Yahwehism. Coming from the position that the male delivered seed into the 'helpmeet', woman, this was reproductive anarchy! There were children who couldn't identify their fathers; and fathers who didn't know they had children. Men were delivering their precious seed, a part of God and infused with his spirit, into the hands of the brazen enemy. It was baby theft, and totally unacceptable.

The Levite priests had to have a story which illustrated the need to turn away from the goddess religion and hearken to the word of God, and it had to be in a language the people understood. Enter Adam and Eve. As assimilation is one of the best ways to 'disappear' the opposition, they adopted elements of the earlier Sumerian and Babylonian creation myths, and incorporated them into the new version of the 'truth'. The idea that God created man from dust was borrowed from the Sumerian city of Nippur, whose theology had it that mankind (man and woman) had been created from clay by the goddess.

In the Sumerian city of Erech, a tablet was found which described a young maiden, Lilith, as the 'hand of Inanna'. Her job was to go out on to the street and invite men back to Inanna's temple.

However, by the time the handmaiden of the goddess became a Jewish legend, she had become a baby-killer and Adam's first wife. Louis Ginzberg tells the story:

> To banish his loneliness, Lilith was first given to Adam as wife. Like him she had been created out of the dust of the ground. But she remained with him only a short time, because she insisted upon enjoying full equality with her husband. She derived her rights from their identical origin. Lilith flew away from Adam, and vanished into the air. Adam complained before God that the wife He had given him had deserted him, and God sent forth three angels to capture her. They found her in the Red Sea, and they sought to make her go back with the threat that, unless she went, she would lose a hundred of her demon children daily by death. She takes her revenge by injuring babes – baby boys during the first night of their life, while baby girls are exposed to her wicked designs until they are twenty days old.[4]

Lilith, the woman who demanded equality, became the very antithesis of what woman's role in the male-seed culture was to become: obedient nurturer of men's children. Babies wore amulets to protect them against Lilith. She attacked men's reproductive potential in other ways too. In the Kabbalah, compiled in the sixteenth century A.D., Lilith was called the 'Queen of the Demons' and blamed for arousing men's sexual impulse when the proper recipient, the wife, wasn't around. Masturbation would provide the seed Lilith needed to make her demon-babies and thus became a highly dangerous activity. Nocturnal emissions were explained as Lilith up to her nefarious ends.

According to Ginzberg, Adam had been repelled by Lilith because he knew 'well all the details of her formation', she having been made 'in his presence', from the dust of the ground. God wouldn't make this mistake twice. Apparently, 'only when like is joined unto like the union is indissoluble' so, on the second try, the wife would be taken from Adam himself. First, though, Adam was made to fall into a deep sleep 'for, had he watched her creation, she would not have awakened love in him.'

Ginzberg explains the reason for choosing the rib:

> When God was on the point of making Eve, He said: 'I will not make her from the head of man, lest she carry her head

high in arrogant pride; not from the eye, lest she be wanton-eyed; not from the ear, lest she be an eavesdropper; not from the neck, lest she be insolent; not from the mouth, lest she be a tattler; not from the heart, lest she be inclined to envy; not from the hand, lest she be a meddler; not from the foot, lest she be a gadabout. I will form her from a chaste portion of the body' and to every limb and organ as He formed it, God said 'Be chaste! Be chaste!' Nevertheless, in spite of the great caution used, woman has all the faults God tried to obviate.[5]

Amaury de Riencourt has another explanation for the choice of the rib in the creation of Eve story: 'Even the fashioning of Eve out of one of Adam's ribs, a symbol of man's priority, is an adaptation of a Sumerian play on words – the Sumerian for rib, *ti*, also means 'to make live'.[6]

Gerda Lerner makes this point rather differently, but first explains the context in which the rib became important:

The description of the Garden of Eden parallels the Sumerian garden of creation, which is also described as a place bordered by four great rivers. In the Sumerian creation myth, Mother-Goddess Ninhursag allowed eight lovely plants to sprout in the garden, but the gods were forbidden to eat from them. Still, the water-god Enki ate from them, and Ninhursag condemned him to die. Accordingly, eight of Enki's organs fell ill. The fox appealed on his behalf, and the Goddess agreed to commute sentence of death. She created a special healing deity for each afflicted organ. When it came to the rib, she said: 'To the goddess Ninti I have given birth for you.' In Sumerian the word 'Ninti' has a double meaning, namely, 'female ruler of the rib' and 'female ruler of life'. In Hebrew the word 'Hawwa' (Eve) means 'she who creates life', which suggests that there may be a fusion of the Sumerian Ninti with the Biblical Eve.[7]

The word 'Ninti' is thus seen as the joining of '*Nin*' – 'female ruler' (as in Nin-hursag) – and *ti*, as de Riencourt says, meaning both 'rib' and 'to make live'; or, as Lerner says, 'female ruler-of the rib' or 'female ruler-of life'.

With the benefit of the Sumerian myth, we can see the similarities and differences with the Jewish version. The setting of the Garden of Eden is similar, bordered by four rivers. Both have forbidden fruit but in the Sumerian myth it's the god Enki, not Eve, who challenges the divine order. Enki was the god of water who became important after the advent of irrigation, and could have represented the idea of male stimulation of female seed. In Sumeria, he was the cheeky upstart, not Eve. The last obvious difference between the Sumerian and biblical Garden of Eden myths, of course, is that in the Sumerian version it was created by a female while the biblical one was created by a male.

Jewish scholar Louis Ginzberg says that after Adam and Eve had eaten the forbidden fruit, none of the trees would allow Adam to take leaves from them so he could cover his nakedness. Except one: 'Only the fig-tree granted him permission to take of its leaves. This was because the fig was the forbidden fruit itself.'[8] This was probably *ficus sycomorus*, a tree much favoured in the Middle East because its large size and huge heart-shaped leaves provide excellent shade. The figs themselves grow in clusters, rather like grapes. On ancient Egyptian murals the goddess Hathor was shown within the tree, handing out the fruit, according to Merlin Stone, 'to the dead as the food of eternity, immortality and continued life, even after death'.[9]

Whether or not the sycamore fig was actually consumed during the rituals of the goddess in the lands the Hebrews visited and settled, it's very likely that the worshippers did meet under its shade. In any event, they certainly met in 'groves' of trees, as the Bible continuously points out. This was perhaps because, as we can see in Sumerian artefacts, the goddess had long been associated with trees.

There is universal agreement among scholars that the serpent in the Adam and Eve story is a direct representation of the goddess. From earliest times in Europe and elsewhere, images of the goddess were intertwined with snakes. The snake or serpent was associated with the goddesses Ua Zit, Hathor, Nibada, Ninlil, Inanna (Nina), later Ishtar, Tiamat, Athena, Hera, Gaia and others. They appeared together in Egypt, Crete, Cyprus, Babylon, Canaan and elsewhere. It was therefore inevitable that the serpent should be chosen as the being which enticed Eve into disobeying the word of God.

Ginzberg tells us that the serpent's poison was 'the poison of evil inclination' and that after the serpent had plucked the fruit for Eve she 'opened the gate of Paradise, and he slipped in'.[10] These euphemisms for sexual intercourse are easy enough to draw, given the phallic shape of the serpent and the unrestrained sexuality of goddess-worship. But there was more to the serpent than that. It is quite possible that priestesses were trained to become immune to snake bites, by small doses of venom taken over a period of time, so that when consulted later for divine guidance, they could be covered with dangerous snakes and bitten. Instead of the usual physiological effects of snake bite, including respiratory paralysis and death, the priestess would have an experience akin to that induced by certain hallucinogenics. They had visions and, as has been reported by those bitten by snakes after immunisation, they may have spoken in verse.[11] To a people ignorant of the mechanism of immunisation, let alone the secret training methods of the priestesshood, such displays of apparent immunity to death must have been impressive. No wonder they thought the priestess was in direct communication with the source of life itself: the goddess.

The Levite priests who, with threats, were trying to educate the people into the new male-centred ways, were having a very hard time. The new concepts were obviously proving difficult to impose. We hear time and time again in the Bible that the 'children of Israel' forsook the Lord, and served Baal and Ashtaroth (Judges 2:13). The reluctance of the people to adhere to the new religion was an abiding source of aggravation to the new hierarchy. It necessitated the introduction of a spirit police, as Ginzberg tells us with this story about one King of Israel, Josiah:

> The efforts of the king on behalf of God and His law found no echo with the great majority of the people. Though the king was successful in preventing the worship of idols in public, his subjects knew how to deceive him. Josiah sent out his pious sympathisers to inspect the houses of the people, and he was satisfied with their report, that they had found no idols, not suspecting that the recreant people had fastened half an image on each wing of the doors, so that the inmates faced their household idols as they closed the door upon Josiah's inspectors.[12]

One can understand the reluctance of women to convert to the new religion and laws; they would lose their parental rights. It seems likely that the 'inspectors' were male, but clearly some men were having difficulty accepting their male order. Perhaps they didn't like the terror tactics that came with it or perhaps, like the crowd at Pathros, they remembered a time when they had 'plenty of victuals, and were well, and saw no evil' (Jeremiah 44:17).

Or perhaps they were just hedging their spiritual bets, as people tend to do.

But loyalty to Yahweh, and fathers in general, had to be accomplished, and the battle went on. It rages on practically every page of the Old Testament, right down to the very last. In Malachi 2:11, we read that 'Judah hath profaned the holiness of the Lord which he loved, and hath married the daughter of a strange god.' This 'god' was, of course, the goddess. But, apparently, salvation was in sight. It was to come in the form of the prophet Elijah who, the very last words of the Old Testament tell us, 'shall turn the heart of the fathers to the children, and the heart of the children to their fathers, lest I come and smite the earth with a curse' (Malachi 4:6).

It doesn't say that children's hearts shall turn to their parents (let alone mothers) because general family loyalty is not what this threat is about. Matrilineal descent was the evil the Old Testament was obsessed with and the final message was this: recognise that the children belong to you, not her; and children, recognise where your allegiance belongs – to your father. The 'heart', loyalty, had to be turned away from mother, and away from the goddess. In time, of course, it was.

12 The fall

The difference between the idea of a female-centred creativity and a male-centred one is simply that – an idea. Nobody could prove the point either way until we had a microscope, and even then it was none too clear. The only obvious fact was that men contributed something to reproduction, but what was it? And what was it in terms of (a) physical body and (b) spirit?

To the people who made the figurines and built the temples on the Mediterranean islands of Malta and Gozo between 5000 and 3000 B.C., generativity and spirituality seemed to be encapsulated in the female form. To get close to spirituality one actually entered the body-like enclosure of the goddess' temple through the vagina-like door. The phallus was part of the life process but it was peripheral, stimulatory perhaps. All this had changed by the third century B.C. when huge stone phalluses were placed on columns forming the avenue dedicated to the god of fertility, Priapus, on the Greek island of Delos. Generativity and spirituality were again one – only this time it was male.

Something had happened to change things, but what was it? Perhaps when men took agriculture over from women, with the invention of the plough around 3000 B.C., they began to see themselves as the breakers of earth and sowers of seed, rather than mere irrigators or stimulators of female seed. Alternatively, when some people began to live exclusively pastoral lives, on the vast Eurasian steppes for example, perhaps they became acutely aware of the generative importance of the ram, the male.

Any man could have thought up the idea of a male reproductive seed, and no doubt many did. But the new reproductive theory wouldn't have been much heeded in societies which had for millennia thought in terms of a female creativity; where women were the source of seed, where the goddess was supreme, and priestesses had access to the divine powers, where men lived in their wife's house, surrounded by her kin. To change reproduction theory one had to change an entire way of life, and that wasn't going to happen overnight.

If reproduction were just a matter of physiology, it has always been possible to come to the conclusion there is equal parental input. This could be done using the red–white reproduction theory, many versions of which exist all over the world. This draws on the fact that semen is white and menstruation is red, and says men contribute the white parts – bones, brain, eyes, spinal cord and nervous system – while women contribute the blood and flesh. Depending on the role of spirit, this could be a nice, co-operative venture. If, for example, you say the spirit comes from (or is) an ancestral spirit, it could pass through the physical fusion of male and female parts, so the process is non-sexist, and no superiority complex can develop. But if a man or woman says 'The spirit or soul passes through me alone', through the 'red' or the 'white', then that gives them added importance, responsibility, authority, kudos and prestige. At least it does if other people believe them.

Given the fact that reproduction only starts when men provide the vital input, there was plenty of scope for attributing to men that which made life spring into action – the spark of life, spirit itself. From this, biological logic could easily follow. In the sixth century B.C., Pythagorus had the theory that the soul and the 'moist vapour' from which an embryo would develop came from the brain and nerves of men, while women contributed the 'grosser parts': the blood and humours of the womb. This is a form of red–white theory, but greater value is attributed to the man's contribution. (And, on a slightly different tack, anyone who thinks the child's 'white' brain is made from the man's 'white' semen, and thinks the brain is the seat of the soul, is going to attribute the soul of the child, its life, to the male.)

As a way in to reproductive contribution, 'soul' and 'spirit' have the distinct advantage of being invisible, and once you've got your views established, nobody can prove you wrong. Spirituality might have been the wedge men used to break women's monopoly, but once they had done so, they held on to it. So it is we have Aristotle in the fourth century B.C. talking about men's active seminal power being more 'divine', and even that classic ovist, William Harvey, in the seventeenth century A.D., was still insisting that men contributed an 'incorporeal' (that is, invisible) 'fructifying power'. He wouldn't suggest *women* contributed the 'fructifying power' because by this time women's invisible, spiritual input had been well and truly elbowed out.

Where men had a toe-hold in reproduction, it was probably through a spiritual link. So it still is among the Ashanti of Ghana, who have a strong matrilineal tradition behind them. Inheritance is through the female line and there is, according to Dr Peter Sarpong, a 'general precedence of girls over boys in the estimation especially of Ashanti women'.[1] The land is owned and worked by the women while the men are the 'front' of society; in the old days they were the kings and warriors, now they are the generals, government ministers and civil servants. Sarpong writes: 'In contrast to the mother–child bond which the Ashanti consider to be a biological one, the father–child tie is regarded as spiritual.'[2] The child is actually made up of three elements: the mother's blood; the *kra* or life principle which the Creator gives to a baby just before it is born, 'a small indestructible part of the Creator', as Sarpong puts it; and the *sunsum*, the spiritual inheritance from the father which 'is thought to mould the child's individual personality and character', and which links the child with the father's *ntoro* or 'spirit washing' group.[3] Men are thus drawn into the process of reproduction but, because the biological link is entirely between the mother and child, and because the matrilineal system is so strong, the father actually has little say in the bringing up of the child known by all to be 'his'. According to Sarpong: 'Jural rights over a child lie with its matrilineage members. But as it is believed that by reason of the spiritual bond, dissatisfaction on the part of its father could be fatal to his child, conflicts between him and the child's matrilineage members are kept to a minimum.'[4]

Sarpong explains why, among the Ashanti, 'the value of girls is almost inestimable':

> In them the matrilineage puts the hope for its future existence. It depends upon them to provide suitable persons to take up offices and to strengthen the lineage. Whereas the boy is completely incapable of providing successors for his matrilineage, in the girl the lineage has potential males as well as further potential females.[5]

The contemporary Ashanti are similar to the ancient matriarchies in that they are matrilineal agriculturalists. Women own and work the land, which gives them independence; while names and property are inherited through the female line and the birth of girls is especially welcomed. Like the ancient matriarchies, the Ashanti used to worship a goddess, Nyame, who at a later date became a male god. Nyame, the god, had to share his power with

Asase Ya, the earth goddess who was 'queen mother' of the gods. This relationship was reflected in the Ashante king/queen-mother relationship where she was the ultimate source of power. Likewise, in ancient matriarchies, the source of power was ultimately female.

The ancient matriarchies recognised a male role in reproduction but, as we can see from the Ashanti, that role did not have to be perceived as biological. Men are connected to their children by their *ntoro*, through which a child acquires its individual personality and character. Such a concept weaves the two parental families together, and some variations on this theme may have pertained in other, older places. And as with the Ashanti, such a father–child link need not lead to paternal proprietorial rights.

Because it is important an Ashanti child knows its *ntoro* group, pregnancies are kept identifiable by the practice of monogamy or serial monogamy. This could have happened elsewhere, and didn't mean that women were locked into unsatisfactory marriages. According to Sarpong, Ashanti men: 'know quite well that any ill-treatment of their wives will bring the weight of their matrilineage down upon them to demand either better treatment or divorce'.[6]

Ashanti women are not only financially independent, they are free from male control. Although they may have children by several men, Ashanti women are not promiscuous. In this, we may have a difference between the Ashanti and the ancient matriarchies which practised the 'sacred sexual customs' and the 'sacred marriage'. At least, we can see the difference now, but what happened among the Ashanti many thousands of years ago? Perhaps the *ntoro* group hadn't been thought up yet, and paternity was unimportant. The only clue we have on this subject is the fact that the 'sacred marriage' – the disposal of the king – was practised among African tribes, as it was among tribes all over the early 'civilised' world. Barbara Walker fills in some details: 'Up to 1810 A.D., kings of Zimbabwe were ceremonially strangled to death by their wives at the moon temple every four years ... Nigerian kings were strangled after the queen's pregnancy was established, which meant each king fulfilled his role in life by begetting one royal offspring.'[7]

This 'begetting', however, is not so crucial that the king is going to be allowed to watch his offspring grow up. His job is done and his connection has gone and, as if to highlight the point, he is dispensed with. Alice Schlegal has surveyed contemporary matrilineal societies and compared the position of women in them with their patrilineal neighbours. She writes:

> I hypothesise that the descent system, by emphasising one parent, both parents equally, or each parent in a different way, conditions the attitude that the individual learns toward sexual differences and carries over in his behaviour toward members of the two sexes ... It is probable that the importance of the woman as the linking factor in the descent group gives to womanhood a dignity that may be lacking in societies which do not have this belief. In psychological terms, this is saying that in matrilineal societies there is a cognitive set toward the importance of women that has an effect in mitigating male dominance.[8]

But what happened when there was no male dominance to mitigate, when there were no patrilineal neighbours for people to get ideas from? I imagine the 'cognitive set toward the importance of women' would be rather high.

In *Women in Prehistory*, Margaret Ehrenberg tells us that: 'Many aspects of the archaeology of the Linear Pottery Culture of early Neolithic Europe ... suggest that its social organisation and economic base may have allowed women to be highly valued and play a leading role in many aspects of life, enjoying status *at least equal* to that of men,' (my italics).[9] So what were the position and status of men? Some feminist commentators have tried to stress the point that there was equality between the genders. In *The Women's History of the World*, for example, Rosalind Miles says that ancient matriarchies were 'woman-centred' but 'substantially egalitarian'.[10] A few pages later she tells us that 'Every myth, every song in praise of the Great Goddess stressed by contrast the littleness of man, often in caustically satiric terms.'[11] If this is the case, men might have felt they weren't quite so equal. Marija Gimbutas says that the evidence points to a 'balanced, nonpatriarchal and nonmatriarchal social system',[12] but James Mellaart said of the Anatolian town of Çatalhöyük that men were 'utterly subject to their mothers'.

It is, of course, difficult to know the degree of self-esteem men and women experienced in prehistoric times but one thing is certain: a female-centred reproductive theory does not lead to domination of the male. With the female-centred life-view a woman might understand that she needs a man to stimulate her reproductive potential, but she needs him for a very short time, long enough to make love to her. Thereafter, she is reproductively independent; she carries the baby within her and gives birth to her child herself. On the other hand, a male-centred reproductive theory causes a profound void to appear between the source of generativity, the male, and the means of production, the female. To bridge that gap, and ensure paternity, men needed to control the sexuality of women. Men could never be reproductively independent, as women once were, and with the new male-centred reproductive theory, the notion of control – one human being having the right to control another – took a quantum leap forward. This is why, when trying to figure out the nature of ancient matriarchies, we should not assume they were a mirror-image of patriarchy, a nightmare the other way around.

The battle

The male-seed/female-soil reproductive theory caused a psychological revolution which can be seen in mythology. Amaury de Riencourt writes:

> It is in the amazing metamorphosis of the old matriarchal mythologies that the depth and importance of this profound psychological revolution can be best understood. All the mythologies were taken over and stood on their head; all the female-oriented myths were reinterpreted patriarchally. What had been good became bad, former heroes became demons, and the remarkable coincidence was that this metamorphosis happened more or less simultaneously, over a period of time, in the Indo-European and Semitic worlds, in Greece as in India, and in China as well as in Palestine and Mesopotamia. It is to this exceptional conjunction of a changing male outlook in the ancient agricultural societies and an irresistible tidal wave of nomadic patriarchal invaders that history owes this global revolution.[13]

The goddess was tumbled. Where, once, women had been priestesses, they were now barred from religious office. Where the family name and land had passed through the mother, it now passed through the father. Where women had been head of the household, now men were. And where women had been independent, they were now obedient wives, controlled by men.

It's a matter of intense academic debate as to who exactly these 'northern invaders' were who wreaked havoc upon the burgeoning civilisations of Europe and the Middle East. We do have details of some of the invasions and can trace the movement of, for example, the Kurgans, Hittites, Hurrians and Kassites. But tracing these people back to their original homelands is made difficult by the fact that they appear to have had few cultural centres of their own. With such confusion, it is perhaps inevitable that they get called by generic terms such as the 'battle axe cultures', Indo-Europeans, Indo-Aryans or Aryans. What allows us to so classify these people is that they had certain things in common: they were male-ruled and had male gods; they were pastoral – they herded animals; and they carried a great deal of weaponry which they used against other people in other lands.

According to Marija Gimbutas, Old European culture, with its 'striking absence of images of warfare and male domination',[14] was brought to an end by repeated invasions by the Kurgans, 'changing it from gylanic to androcratic and from matrilineal to patrilineal'. This happened between 4300 and 2800 B.C., with western Europe and the Aegean and Mediterranean regions escaping the disturbances the longest, while the Mediterranean islands managed to maintain their peaceful and creative lifestyle until 1500 B.C. She describes the Kurgans thus:

> The basic features of the Kurgan culture go back to the 7th and 6th millennia B.C. in the middle and lower Volga basin [in south Russia] – patriarchy; patrilineality; small-scale agriculture and animal husbandry, including the domestication of the horse not later than the 6th millennium; the eminent place of the horse in cult; and, of great importance, armaments – bow and arrow, spear, and dagger.[15]

The question is, of course, why should the Kurgans and other 'battle axe cultures' have developed a patriarchal outlook while others living further to the south were content with their matriarchies? I believe the answer lies with differing ideas about reproduction, ideas which are reflected in religion, myth, iconography and lifestyle. At some indeterminate time in the very ancient past the Kurgans had taken a momentous leap into virgin territory – the vast Eurasian steppes. Following their grazing herds, the Kurgans had left whatever tradition they then had, possibly a female-centred one, far behind. The animals now became of paramount importance and, with little else to do on that bare and open terrain, the pastoralists watched them day and night. Over time, the ram came to be seen in a new light: perhaps he wasn't stimulating life in the female but actually delivering it. The idea appealed to the men while the women, isolated from matrilineal traditions, couldn't prove them wrong. As the older women died, the memories of another era died with them. Eventually, the patriarchal idea took hold.

From this base, north of the Black Sea, Caucasian Mountains and Caspian Sea, the Kurgans later migrated. Why they should have done this is, like almost everything else to do with their history, a matter of debate. Some think the crucial factor was their ability to move fast – they had domesticated the horse. Others say they were just warmongers, as evidenced by their love of weaponry, including jewel-encrusted battle axes; while others say their skills in metallurgy turned them into warmongers; and there are those who say there was a population explosion. While some or all of these theories may have contributed to the Kurgan phenomenon, I believe they were driven by an abhorrence of matrilineal descent patterns.

When the Kurgans finally made their way out of their patriarchal stronghold, they came upon a world of madness. Everything was upside down and back-to-front. They found people who didn't think in terms of male seed, of children begotten from many generations of men and, ultimately, a male god. Instead, they found men without children, and women who claimed children as their own – men were peripheral to life, not its driving force. To the Kurgans, it was a world in which the progeny were lost, wandering around ignorant of their roots. It must have seemed an abomination, and a profound insult to their male gods and, with might on the side of right, the Kurgans set off on their crusades to put the world straight.

Although, on a global scale, the male-seed revolution took a few thousand years to accomplish, on a local level the conversion may have been swift. Small communities could be dealt with by killing all the men and male children, thus getting rid of the enemy's seed; by killing the older women, and with them the alternative life-view; and by retaining only the younger women and girls who could be used as baby-making machines for the invaders.

In other places, where the relative numbers didn't allow such drastic measures, the invaders had to bide their time, promote their 'new biology', take over the power-base shrine by shrine – all the while watching the old women die and, with them, knowledge of the old ways.

Perhaps some men living in women-centred societies were pleased when they saw the invaders thundering over the horizon on their horses, and embraced the idea of man-the-creator with enthusiasm. Certainly, there would have been many for whom the male-seed concept came as a very nasty shock – they suddenly realised they didn't know their roots. All the time they had been tracing their lineage through their mother's line, they should have been tracing it through their father. And many men would not have known who their father was – a loss they must have felt deeply. Angry and cheated, there must have been men for whom the new biology made perfect sense and who were only too willing to accept the new god and rule on the invader's behalf. And despite the horror of forcible invasions, for all men there was the compensation that came with the new biology: men made life and the children now 'belonged' to them.

Some places were hard to conquer. An example is the ancient city of Alalakh, on the Turkey–Syria border, excavated by Sir Leonard Woolley. Around 2300 B.C., invading nomads captured the city and converted the goddess's shrines to those of gods. Heads of bulls, symbols of the goddess, were replaced by heads of rams, symbols of patriarchy. In 1850 B.C., however, the people rose up against 'the hated kingdom of the conquerors', as Woolley put it, and burnt the king's palace and temple to the ground. In 1200 B.C., there was another about-turn. In the forecourt of the temple Woolley found 'a much defaced limestone figure of a seated goddess' and 'a smashed bull image'. The only object left in one piece was a white limestone ram's head – the patriarchs had won the final battle.

Individual men could have heard of the new biology as they travelled and traded, or had men visit their area, carrying the new idea. Some would have demanded a new order at home and been told by the women to 'get lost' if they didn't like the female-centred traditions. Other men stayed and fought. One such character was Rama, of the *Ramayana* fame who, according to Fabre d'Olivet, proved unable to overthrow the matriarchy in his homeland – somewhere in Anatolia or southern Europe – and, with his band of followers, went East looking for new lands to impose his ideas upon.[16] About 3000 B.C., he arrived in India and 'won' the hand of the hereditary princess, Sita. He then proceeded to treat her viciously, soon establishing himself as the monarch. On his death,

India was thrown into centuries of war between his followers, the Pandavas, and the followers of the ancient goddess, the Kourava. The Pandavas won.

Similar dynamics are evident from innumerable ancient texts. For example, the Sumerian god, Nergal, 'won' the hand of the Goddess of the Underworld by dragging her from her throne by her hair, and threatening to kill her. The patriarchal revolution was won by violence, not love.

The earliest known statement of people's rights comes to us from the Sumerian city of Lagash where, in 2300 B.C., Urukagina inscribed his 'reforms' into the walls of buildings. They mention that 'women of former times each married two men, but the women of today have been made to give up this crime'. This statement has generated much comment and is open to interpretation. However, Professor Saggs, an expert on Mesopotamia, writes: 'the reforms of Urukagina refer to women who had taken more than one husband; some scholars shied away from this conclusion suggesting that the reference might be only to the remarriage of a widow but the wording of the Sumerian text does not really support this'.[17]

The Reforms of Urukagina are unique as a historical record of a Sumerian city and may not be typical of what happened. However, it's perhaps worth noting here that they tell the story of how, 50 years earlier, in 2350 B.C., a man called Lugalanda had overtaken the temples and usurped their power. He made himself and his wife owners of the temples, their land and flocks, cut the names of the deities out of the temple records and put his own administrators in charge. This was the man Urukagina forcefully overthrew, complaining that he had stolen the deities' property. Throughout, the Reforms of Urukagina are referred to as *amargi*, which has the double translation of 'freedom' and 'return to the mother', implying that by getting rid of Lugalanda the society could return to the old ways. Although, of course, not quite the old ways, as the reform concerning a woman taking two husbands shows. Indeed, the reforms are basically anti-female because at this time Sumer was undergoing the patriarchal revolution, with goddesses being replaced by gods and priestesses by priests. However, the old ways were obviously different because one reform was reinstating the custom of allowing the produce of the temple lands to be made freely available to the needy. This raises the possibility that in the earliest days of Sumerian culture society could have been more communal.

To say these were dynamic times in the Middle East would be an understatement. Waves of patriarchal invaders were changing the power balance and there was inter-city rivalry. Because the record is so scanty, we have only glimpses of the older ways. In the Sumerian state of Eshnumma, for example, rapists were being put to death as late as 2000 B.C. (Whereas in 1250 B.C. the Levite priests were punishing the Hebrew *victims* of rape with death.) A women who had a child by another man while her husband was away at war was not punished for adultery and, in law, was still considered the wife of the first man. But, 'if a man rejects his wife after she bears a child and takes another wife, he shall be driven from the house and from whatever he owns'.

In Egypt, the most ancient deities were Ua Zit, cobra goddess of the north, and Nekhebt, vulture goddess of the south. Their supremacy was overthrown by Horus around 3100 B.C., when the 'Shemsu Hor' (people of Hor), as the Egyptians called them, invaded the country. (The name 'Horus' was the Greek word for 'Hor'.) This period of Egyptian history is shrouded in mystery but some experts believe that the Shemsu Hor were followers of Hor-Aha, known as Menes or Narmer, famous for being the man who united north and south and made Egypt a single country (although this man may, in fact, have been two men). There is plenty of evidence from this period of Mesopotamian or perhaps Syrian contact and some believe that Hor-Aha was a Sumerian who'd been ousted from his homeland by even more patriarchal invaders. In any case, although his unification of the country is often described as a great accomplishment, for the people he conquered it was no picnic, as a picture of him clubbing the locals shows. This appears on the famous 'pallette of Narmer' which reads in pictorial form: 'Pharaoh the incarnation of the hawk-god Horus, with his strong right arm leads captive the Marsh dwellers.'[18] This was the beginning of kingship in Egypt.

The battle of Horus over the native goddesses is told in the story of Hor slaying his brother/uncle Set, the word 'set' meaning 'queen' or 'princess' in Egyptian. The choice of the name 'Set' as the enemy Horus had to overcome may also have been related to the fact that Zet was the serpent of the goddess Gaia, or because Ua Zit was the ancient cobra goddess. In any case, although the new ruling elite worshipped Horus the sky god and, later, Re the sun god, the ancient idea that the deity resided in the earth, to which the dead descended, was never really superseded in the minds of ordinary people. Perhaps this is why the powerful and enduring female force was recognised by the ruling elite who incorporated the ancient symbols of the two goddesses into their crowns, and why the feminine influence was never squashed to the extent that it was elsewhere. Throughout Egyptian history there were plenty of goddesses for people to appeal to including Neith, Au Set (Isis), Hathor, Nut, Tefnut, Sakhmet and Selket.

Egypt was less prone to turmoil because it was geographically less accessible although, in terms of female independence, it certainly had its ups and downs. During the Old Kingdom, between 2686 and 2160 B.C., women held important positions as priestesses, administrators and so on. Then there was the First Intermediate Period – 120 years of social and economic instability caused by climatic changes – and the Middle Kingdom, during which women were obviously out of favour because they ceased to hold positions of power and appear in the records in 'helper' roles such as scribes, hairdressers, gardeners, millers and so on. Again, there was an 'Intermediate Period' – more turmoil caused by the invasion of a patriarchal Asiatic peoples called the Hyskos – until order was restored by the Theban forces, when Egypt entered 'The New Kingdom', a highly organised and prosperous time lasting from 1567 to 1085 B.C. We have numerous texts from this period from which it is quite clear that women had a legal status equal to that of men and lived independent, active lives, while sculptures and other visual artefacts, showing kings and queens holding their arms around each other in a relaxed and happy demeanour, indicate that affection, not control, was what kept them together.

By the time the Greek historian Diodorus Siculus arrived in Egypt in the first century B.C., the male-seed theory of reproduction had taken hold. He wrote 'the Egyptians hold the father alone to be the author of generation, and the mother only to provide a nidus and nourishment for the foetus'. Whether the idea had been imported by the invading Hyskos, Hittites, Assyrians or by Alexander the Great is unknown; it could have been any of them and no doubt they compounded each other's influence in this matter. Yet, through it all, the traditional matrilineal system of inheritance prevailed because, as Diodorus recorded, 'only the daughters inherit in Egypt'.

Although this seems paradoxical, we have to remember that in the switch from matriarchy to patriarchy, which took over 3000 years in Egypt, there is an interface period which is bound to be full of paradox. Indeed, as we ourselves move away from patriarchy, we can see paradox in our own lives. Women are 'liberated' in that they can take almost any job they're capable of, but invariably still have the 'helpmate' role at home, washing, cooking, cleaning and so on. Invading armies, moreover, do not always exert a deep influence on the traditional ways of the people they have under their military control. Nor do 'official' religions necessarily impress the general populace, or affect traditional customs.

Today, we can see innumerable examples of societies living a 'double life'. Matrilineal descent patterns exist side-by-side with Islam among the Minangkabau of Indonesia, and beside powerful brother-right among the Yoruba of Nigeria. Such anomalies are bound to occur and there will be many permutations within the general theme of transition. Interestingly, the back-and-forth nature of the matriarchy–patriarchy battle has been recorded by Nancy Tanner amongst the Minangkabau. During the Dutch colonial period the nature of the matrilineal customs changed so that the mother's brothers became controlling but today, some 80 or so years later, there is a much greater degree of female autonomy as the people have returned to the traditional ways, with the mothers themselves in control.[19] The matrilineal descent system has this elasticity, and no doubt Egyptian women experienced it as patriarchal influence waxed and waned.

You'd think, perhaps, that once men had tasted power they would never allow a return to the female-centred ways. But we can see from the Minangkabau that this isn't so. Although the brothers of the mothers were once in control of the family, they are no longer. In 1990 Richard Mahler visited these people living in Sumatra and he reports that 'within the family, the eldest woman is responsible for all important decisions affecting the group, although she may often consult with other female family members before making those decisions'.[20] Apparently, this allocation of power doesn't bother the men:

> In my conversations with Minangkabau men, they seemed to be relatively content with the situation and often turned over all their income to their wives, who gave them a small allowance or pocket money ... Several men told me that when they were young they went off to live on Java or Bali to get out from under the presumed yoke of female dominance, only to return to the Minangkabau area because they found they preferred their original way of life after all.[21]

The Minangkabau are Muslims and pray five times a day and observe the other practices of Islam yet, according to Mahler, 'they see no conflict with their Minangkabau customs'. This proves the point that other matriarchal societies could well have been operating in areas which history has designated 'patriarchal' because a male-centred religion is known to have been the official religion at the time. Islam arrived in Indonesia only in the sixteenth century, and probably reached the Minangkabau some time after that, but their own culture is thousands of years older and, unlike most other Muslims, the Minangkabau value girls more than boys:

> The birth of a daughter is cause for celebration, and a pair of sharp peaks is added to the steeply pitched roof to signal the blessed event. In fact, travellers can determine the number of girls in a family by counting the points on a Minangkabau house and dividing by two.[22]

Today, some people are still matrilineal but clearly patriarchal, like the Yoruba of south-western Nigeria. Although the women own the land, they are dictated to by their uncles and other male relatives, who use women as pawns in their power games. A baby girl can be betrothed to a man 30 years her senior, and no doubt the uncle (the mother's brother) who arranged the marriage will receive a nice fat 'gift' for his trouble. Perhaps the difference between the Ghanain Ashanti and Nigerian Yoruba experiences is that the Yoruba were invaded in 1800 by the Muslim Hausa to the north, and assimilated their patriarchal ideas.

However, as we have seen from the Minangkabau, the switch from matriarchy to patriarchy is not dependent simply upon religion. Force comes into the subject, certainly, but behind force there is always an idea. Differences in cultural patterns may also be due to the skills of a particular orator, on either side, at any particular time. A breakdown in the old ways may be due to a loss of faith in the old female shamanistic powers due to violent invasion, or to negative natural forces such as prolonged drought. Many factors are involved in a takeover – religious, political and social – and when looking at the male–female power balance, each culture must be seen as unique.

Portrait of a cover-up: American Indians

Who would have guessed from watching all those cowboy and Indian movies that when the pale-faces arrived on the American continent the Indians worshipped a female deity, considered women the 'owners' of the land, and were ruled by women? These facts emerge from Paula Gunn Allen's excellent book, *The Sacred Hoop*. It was her experience growing up within the Keres tribe, with the oral tradition still around her, that led Allen to carry out extensive research into early American Indian culture. She writes of 'hundreds of tribes forced into patriarchal modes' and of gynocratic tribes of which little written evidence exists, including the Montagnais-Naskapi, Keres, Navajo, Crow, Hopi, Pomo, Turok, Kiowa and Natchez in North America, and the Bari and Mapiche in South America 'to name just a few'.

We are lucky that Paula Gunn Allen was born at a time when her own culture's roots were just discernible enough to be recovered. In her quest, we can see how patriarchy does its work of 'disappearing' the opposition:

> Women's rituals, ceremonies, traditions, customs, attitudes, values, activities, philosophies, ceremonial and social positions, histories, medicine societies, and shamanistic identities – that is, all the oral tradition that is in every sense and on every level the literature of the tribes – have been largely ignored by folklorists, ethnographers, and literary critics in the field of American Indian studies. These traditions have *never* been described or examined in terms of their proper, that is, woman-focused, context. Actually, it is primarily the context that has been ignored – vanished, disappeared, buried under tons of scholarly materials selected and erected to hide the centrality of women in tribal society, tribal literature, and tribal hearts and minds.[23]

The basic problem is 'the white man's belief in universal male dominance'[24] – a belief which hampers a true understanding of early cultures all over the world, in fact. As in so many other places, the American Indian deities were originally female:

> There is a spirit that pervades everything, that is capable of powerful song and radiant movement, and that moves in and out of the mind. The colors of this spirit are multitudinous, a glowing, pulsing rainbow. Old Spider Woman is one name for this quintessential spirit, and Serpent Woman is another. Corn Woman is one aspect of her, and Earth Woman is another, and what they together have made is called Creation, Earth, creatures, plants and light.[25]

This ancient female Creative Spirit was not simply a 'fertility goddess': she was the 'power of intelligence'; the 'Old Woman who tends the fires of life'; and as Old Spider Woman, she was the one who 'weaves us together in a fabric of interconnection'. Today one can see the last stages in the process of Her being lost – replaced by male-gendered creators, including the generic 'Great Spirit':

> The Hopi goddess Spider Woman has become the masculine Maseo or Tawa, referred to in the masculine, and the Zuñi goddess is on her way to malehood. Changing Woman of the Navajo has contenders for her position, while the Keres Thought Woman trembles on the brink of displacement by her sister-goddess-cum-god Utset. Among the Cherokee, the goddess of the river foam is easily replaced by Thunder in many tales, and the Iroquois divinity Sky Woman now gets her ideas and powers from her dead father or her monstrous grandson.[26]

Annihilating the female Creative Spirit is an essential element of the patriarchal takeover, and always has been. Patriarchy teaches superiority of the male over the female, which engenders disrespect of women, but one can't easily disrespect the woman one lives with if, at the same time, one accepts female primacy as Creative Spirit! In the original Keres creation myth Thought Woman first created the twin sisters Naotsete, who was concerned with internal affairs, and Uretsete, who maintained the psychic and tribal boundaries of the tribe. This is how she did it:

> When Thought Woman brought to life the twin sisters, she did not give birth to them in the biological sense. She sang over the medicine bundles that contained their potentials. With her singing and shaking she infused them with vitality. She gathered the power that she controlled and focused it on those bundles and thus they were 'born'.[27]

Just as Thought Woman's generativity is basically vitalising, rather than simply biologically mechanical, women themselves were credited with having this transformative power:

> The old ones were empowered by their certain knowledge that the power to make life is the source and model for all ritual magic and that no other power can gainsay it. Nor is that power really biological at base; it is the power of ritual

magic, the power of Thought, of Mind, that gives rise to biological organisms as it gives rise to social organizations, material culture, and transformations of all kinds – including hunting, war, healing, spirit communication, rain-making, and all the rest.[28]

According to Allen: 'there is reason to believe that many American Indian tribes thought that the primary potency in the universe was female, and that understanding authorizes all tribal activities, religious or social'.[29]

This helps explain why women were central in the power structure. Allen describes how the Iroquois federal system was made up of local, then 'state', then 'federal' bodies, each of which had executive, legislative and judicial branches. Theological precepts ultimately determined policy at all levels and, significantly, general policy was determined by 'the Council of Matrons', who were 'the ceremonial center of the system'. Judicial decisions were made by men and women together, but the Council of Matrons had a say in which men could join in the decision-making process. Once in the Council, a Matron could not be disempowered unless she herself violated certain laws, one of which was that, if she married, she should not take her husband's name.

Under this system, the women were considered 'the progenitors of the nation, owning the land and the soil'.[30] This shocked the white male invaders, as did the 'petticoat governments', and they conveyed their abhorrence to the Indians. By the early nineteenth century, the old ways had been changed by 'the code of Handsome Lake', who had learnt to resent the 'meddling old women'. Allen writes: 'Handsome Lake advocated that young women cleave to their husbands rather than to their mothers and abandon the clan-mother-controlled longhouse in favor of a patriarchal, nuclear family arrangement.'[31]

Paula Gunn Allen has done feminism a tremendous service in researching the female role in American Indian life and we can only hope that more American Indian women will follow her trail and help us build a clearer picture of America's almost forgotten past. But had the decimation of Indian life and culture been completed by, rather than begun in, say, 1500 A.D., there would probably be almost nothing left for feminists to reclaim today. Imagine, then, how easy it has been for men to 'disappear' female-centred cultures that existed two, five, ten and twenty thousand years ago.

Part Three

The parthenogenetic woman

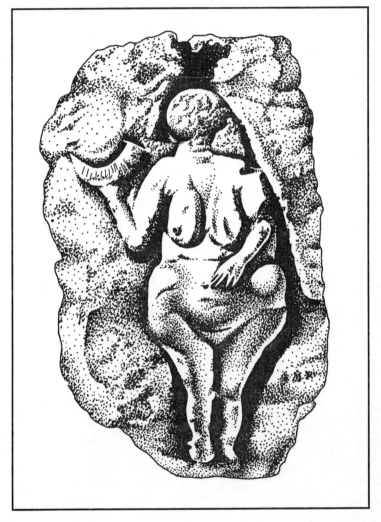

Venus of Laussel
Low-relief in limestone found at rock shelter at Laussel, Dordogne,
France and dated at around 25,000–20,000 B.C. (Musée d'Aquitaine,
Bordeaux, France)

13 The parthenogenetic woman

The first thing people realised about reproduction was that it had something to do with menstruation. Some time between 25,000 and 20,000 B.C., people living in the Dordogne region of France carved a low relief figure of a woman above the entrance to a rock shelter. Known as the Venus of Laussel, the figure, carved in limestone, stands 45 cm high. In her right hand she holds a horn on which there are 13 incisions, while her left hand is placed on her abdomen. As a woman has 13 menstrual cycles a year, this figure could be saying 'menstruation gives women the power to create life'.

There's a cave painting dated at 12,000 B.C. in Ignatievskaya, Russia, which shows a woman with open legs between which there are three lines of dots, totalling 28, the number of days in a menstrual cycle (at least, that is an average figure, but it is the one the medical professions use today). Twenty-eight days times 13 times a year gives the number of days in a year, as close as is possible: 364.

What these numbers tell us is that Paleolithic people, in Europe and Russia at least, understood two things: women cannot have babies until menstruation has begun; and that there is a regular cyclical pattern associated with it. Aside from indicating that menstruation can be used as a measure of time, a calendar, the numbers tell us that the menstrual cycle was then the same as it is now, which in turn tells us that women ovulated, as women do today, for between two to five days a month. That means that then, as now, women could conceive only on between 26 and 65 days a year, and that sex on the other 339 to 300 days would not have produced a baby. For this reason, the connection between intercourse and babies was none too clear and, indeed, was very unlikely to have been made.

In addition, women would have breast-fed for long periods of time, an activity that suppresses certain hormones and makes conception less likely. Paleolithic women might have been having regular sex for between two and four years and without producing any new child. Today among tribal women the connection between breast-feeding and not having babies is well understood, to the extent that a woman will breast-feed a baby monkey, for example, precisely to prevent conception. These contemporary women might understand the connection between intercourse and babies, but in earlier times women may have simply thought that by breast-feeding no babies will come.

Aside from the breast-feeding factor, the connection between cause and effect is easily lost when nine months lay between them, especially when only 26 per cent of the population (say 13 per cent of women) lived beyond age 31,[1] making the gestation period seem, relatively speaking, closer to two of our years. And, when men's facial features were hidden behind bushy beards, I imagine that any resemblance between child and father was less easy to see. With all the factors taken into account it is, then, most likely that our Paleolithic ancestors saw the miracle of life as an entirely female phenomenon. They probably thought one plucked a baby from the body of a woman as one plucked a fruit from the tree or a root from the earth.

Some people find it hard to believe that people once thought women reproduced parthenogenetically (on their own, without men). However, it is interesting that among non-human primate groups paternity appears to be unknown.[2] Female apes just have babies and get on with the business of rearing them without the assistance of males – unless, that is, the males are seeking sexual favours in which case they show what wonderful guys they are by sharing food with the desired female and her offspring.

But, of course, we are not monkeys, and our 'instinct' in such matters might be thought of in terms of a deep spiritual bonding during intercourse. Women tell me, 'but I knew the night I got pregnant, it was special'. Maybe so. However, aside from the fact that conception also occurs when the sex is not special, and the fact that many women are 'sure' they got pregnant when it later transpires they did not, it is physiologically impossible to 'know' that a connection between the ovum and sperm has been made. The winning sperm must wait, with the others, for half an hour near the cervix before continuing its journey. We used to think there was something in the vaginal secretions that 'conditioned' the sperm and allowed them to continue, although it now appears that some chemical in the zona pellucida – the transparent wall that surrounds the ovum – sends a message to certain sperm, allowing them to continue. Once so activated, it takes the sperm from half an hour to six hours to reach the fallopian tube, where it meets the ovum. If the ovum isn't there, the sperm might have to wait around for a day or two. The surety of fertilisation cannot then be known during intercourse and is unlikely even during post-coital bliss. All we can say is that the sperm got a good send-off.

The profound spiritual/magical aspect of womanhood was very much a feature of Paleolithic culture, as evidenced by archaeological finds from France to Russia, and elsewhere. According to Abbé Henri Breuil: 'By far the commonest and probably the earliest art motif is the subtriangular form with a pronounced cut across its narrower end, which has usually been interpreted as a vulvar symbol.'[3] But you can be forgiven for not realising that the vulva was *the* quintessential image of the Paleolithic era because sexist archaeologists prefer to interpret the symbol as an animal hoof-mark (which in fact the vulva may well have been associated with, for reasons we shall later see), while coy archaeologists left these images *in situ* when they removed all the other material evidence to museums in the cities. So, today, you're more likely to find them in the small, local museums near the sites in question where, obviously, fewer people will see them. However, there is no hiding the fact that the Paleolithic was a time when people were exceedingly obsessed with the fecundity of the female – as evidenced not only by the vulva shapes, but by vagina-like fissures in cave walls which have been decorated with red ochre flowing like menstruation to the floor, and breast-like stalactites on cave ceilings carefully decorated with many red ochre dots.

This imagery is not to do with sex, if one defines 'sex' as intercourse. There are certainly no two *or* three-dimensional representations of a penis going in a vagina from the pre-Neolithic eras, and only one image, an engraving on the rock-shelter wall at Laussel, that seems to depict lovers in sexual embrace, although it is so indistinct that it may depict lesbian love as easily as heterosexual love. Indeed, the penis is conspicuous by its virtual absence. Although the caves of southern France and northern Spain have phallic-shaped stalactites and stalacmites, they are not decorated, or altered in any way so they more resemble a penis. They are bypassed as if of not great import. And there are only two examples of three-dimentional 'phalli' (one of which doesn't even look like a phallus), from Abi Blanchard and Belcayre.

Primary sexual characteristics in general were not the concern of
these people – they were not even shown very often on animals.
What interested Paleolithic people was the primary sexual characteristics of *women*: the vulva, menstruation and breasts. When
it came to life and death, the female was the powerful form.

Algerian stone drawing
Paleolithic. Located at Tiout in the Saharan Atlas mountains of North
Africa

As we shall later see, cave art from Africa to India shows ancient
women hunting, and certain contemporary women still hunt, but
Paleolithic women had a magical/spiritual connection with the
hunter – a connection that was made via her vulva and menstruation. (The magical connection between menstruation and
hunting is still maintained among some African hunter-gatherer
groups, including the !Kung, !Xo, /Xam and San, as shown by
Camilla Power in *The Woman with the Zebra's Penis*.)[4] At Tiout in
the Saharan Atlas Mountains in Algeria, one rock engraving shows
a woman with upraised arms, a line extending from her vagina
to the penis of a hunter who has his bow and arrow poised ready
to shoot at an ostrich. At Ignatievskaya in Russia, the central row
of dots coming from the woman's vagina could be extended so
they meet with another row of dots coming from a bison. The
bison and the female are the only two figures painted in red; the
other images – of various exotic animals, some of which seem
deformed – are painted in black. This was no accident. Red, in
the form of red ochre, has been used by innumerable people all
over the world to denote life, death and regeneration.

Before menstruation became 'the curse', it was held sacred. Paula Gunn Allen writes about American Indian Culture, as it once was:

> The water of life, menstrual or postpartum blood, was held sacred. Sacred often means taboo; that is, what is empowered in a ritual sense is not to be touched or approached by any who are weaker than the power itself, lest they suffer negative consequences from contact. The blood of woman was in and of itself infused with the power of Supreme Mind, and so women were held in awe and respect.[5]

According to Stephan Beyer in *The Cult of Tara: Magic and Ritual in Tibet*, Tibetan lamas used menstrual blood in their rituals. Aboriginal Australian myth says that red ochre was created where the female ancestors dropped their menstrual blood. Paleolithic caves show vulva shapes from which menstrual blood, in the form of red ochre, flows to the ground.

Red was the colour of womanhood, of life, of death, and regeneration, and it was found in innumerable ritual contexts. It covered people in their graves, it decorated figurines, egg-shaped boulders and shells. In 4500 B.C. Sardinia, people were buried in the usual ancient foetal position and covered in red ochre. In one grave, next to a female figurine, archaeologists found two halves of an open shell filled with red ochre. In the hypogeum at Hal Saflieni in Malta, an underground complex of many rooms dated between 4000 and 2500 B.C., the excarnated bones (with flesh removed) of between six and seven *thousand* people were found, all amply covered in red ochre. The Venus of Laussel and many other female figurines were originally decorated with red ochre, and whether representative of menstruation, or the blood that pours from a woman when she gives birth, or the placenta, it was a ritualistic requirement. Because of this, ancient peoples often made their homes near deposits of it.

The Paleolithic goddesses

So far, over two hundred Paleolithic human figurines have been found and 99 per cent of them are female. Although they come from a vast geographical area stretching from northern Spain to Siberia, they are remarkably uniform in that they seldom have facial features, although the hair is often elaborately carved; they often have large, pendulous breasts; and they have a great deal of body fat, especially around the buttocks. Although, as a group, they are often described as 'pregnant' (because it fits in with the limiting notion of a 'fertility goddess'), most are clearly not – and their hands lie on their breasts, rather than on their abdomens. As a group, they are awesome and powerful.

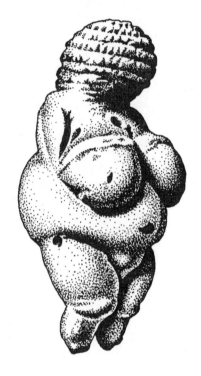

Venus of Willendorf
Limestone figure found near Krems, Austria. Dated between 30,000–27,000 B.C. (Naturhistorisches Museum, Vienna, Austria)

In 1981, anthropologist Patricia C. Rice assessed 188 statuettes from the Gravettian period – about 27–20,000 B.C. – for characteristics which could identify the age of the women they were meant to represent. Having dismissed 56 as too damaged or unclear to classify, she came to the conclusion that the remaining 132 represent not motherhood, but *womanhood*. To ascertain the age of the women depicted – classified as pre-productive, reproductive and post-reproductive – she compared their breasts (high and firm; enlarged and full/elongated due to milk-producing; and sagging), their stomachs (round and flat; protruding/fleshy; sagging), their hips (round and firm; fleshy; sagging), and their faces (unlined; lined; sagging and wrinkled). Then she compared her results with the ages of actual women living in a contemporary hunter-gatherer situation, collected by A. A. Yengoyan,[6] and found they were remarkably similar. Twenty-three per cent of the figurines were classified as young/under 15 years of age (and 23 per cent of the female hunter-gatherers fell into this age group); 17 per cent were between 15 and 35 and pregnant (17 per cent); 38 per cent were between 15 and 35 and not pregnant (41 per cent); while 22 per cent were classified as old/over 35 years of age (compared to 19 per cent in the hunter-gatherer group). What this tells us is that the *smallest* percentage of figurines represented pregnant women, which strongly militates against the idea that these figurines are simply representative of a 'fertility goddess' who, presumably, would be pregnant. As Rice so pertinently asks 'If fertility were the only concern, why so few statuettes of pregnant women, why are none shown in actual childbirth, and why are none shown nursing babies or with children?'[7]

It is truly staggering that *28,000 years ago* people in Dolni Vestonice in Moravia were making fired clay models of women. This tradition of forming images of women – in clay, stone, bone and gold, among other things – continued for another 24,000 years at least, although the meaning of it undoubtedly changed over that time. The Paleolithic figurines are not just about having babies, they are about the power that comes with womanhood – something inherent in the nature of the person who is capable of creating life, whether she is maiden, mother or crone. We have lost this sense of awe of femininity, not surprisingly after 5,000 years of sexist put-down, and may never regain it since we now know that women are not the sole generant. But in these ancient times they probably thought women produced children on their own, without men, and that gave them an authority, and a link with the spiritual which men could never achieve.

Not everyone has seen it in this way. In the early days of archaeology, sexism was rampant and the Paleolithic 'Venus' figurines were belittled in a variety of ways. In 1930, for example, G.H. Luquet thought they were an early form of pornography and said they represented 'not the generative, but the voluptuous character of women ... the thought of the artist was associated much less to the idea of the eventual prosperity of the social group than to the memory or the imagination of his own sensual satisfactions'.[8] It's not surprising that a man who titles his book *The Art and Religion of Fossil Man* should assume these figures were made by men, but the idea that Paleolithic men liked to have sex with faceless women or women wearing a head covering – the Paleolithic equivalent of a paper bag – is truly ridiculous.

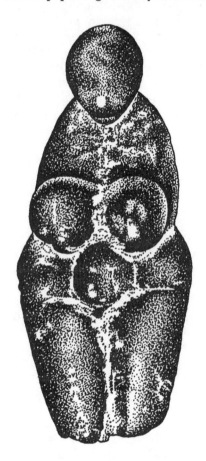

Venus of Menton
Soapstone figurine dated between 30,000 and 27,000 B.C. Also known as the Venus of Balzi Rossi. Found in Ventimiglia, Italy, a few miles from Menton, France, on the Mediterranean coast. (Musée des Antiquités Nationales, St-Germain-en-Laye, Paris, France)

The archaeologist Marija Gimbutas was convinced that these Paleolithic figurines are part of the goddess tradition:

> The beginning of the portrayal of parts of the female body – breasts, buttocks, belly, vulva – goes back to the times when people who did not yet understand the biological process of reproduction (copulation as the cause of pregnancy) created a deity who was a macrocosmic extension of a woman's body. She is a cosmic Creatrix, Life and Birth Giver. These essential parts of the female body were endowed with the miraculous power of procreation.[9]

Concluding her vast and important work, *The Language of the Goddess,* Gimbutas wrote:

> The images and symbols in this volume assert that the parthenogenetic Goddess has been the most persistent feature in the archaeological record of the ancient world. In Europe she ruled throughout the Paleolithic and Neolithic, and in Mediterranean Europe throughout most of the Bronze Age.[10]

Not everyone agrees. In *Women in Prehistory* Margaret Ehrenberg gives two reasons why she thinks 'A universal religion based on a specific female goddess is unlikely in a society such as that of Paleolithic Europe.'[11] The first is 'because it assumes closer and more detailed contact between different groups over a wide area of Europe than is implied by links in other aspects of material culture'. She says 'other' because she has already admitted on the previous page one aspect which is similar – the female figurines: 'Most of the Paleolithic figurines show marked similarities, which strongly suggest a common meaning and linked social or religious tradition throughout Europe.'[12] What other 'material links' does she expect to find from the year 25,000 B.C.? These were not materialistic times! All people had was each other, their natural environment, their spirituality and their language – and linguists are coming to the conclusion that we were once all speaking the same language. Don Ringe of the University of Pennsylvania, says: 'It seems overwhelmingly likely to me that all human languages derive from some common source. I think most linguists would agree with that. I think we would all be shocked if anyone ever came up with hard evidence that all human languages *don't* derive from some common source.'[13]

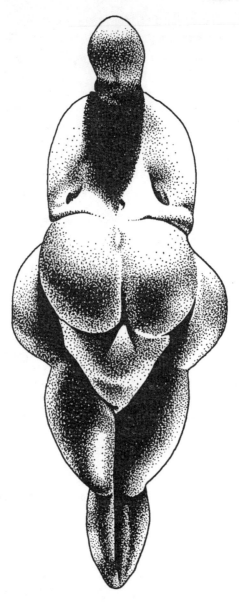

Venus of Lespugue
Carved in mammoth-ivory, and standing 14.7 cm tall, this figurine
has been dated at around 23,000 B.C. It was found at Lespugue, Haute
Garonne, France. (Musée de l'Homme, Paris)

Ringe says this although he is sceptical of those linguists who claim to have *found* a language, Nostratic, from which many ancient languages stemmed. Work on the Nostratic language started in Russia in the 1960s with Vladislav Illich-Svitych, and is continued today by Vitaly Shevoroshkin at the University of Michigan, among others. They claim to have found almost two thousand words which were used 14–15,000 years ago, or even earlier, by a group of people who later developed the Indo-European languages, the Semitic languages, the languages of north Africa and the Altaic languages of central Asia, among others. The interesting thing, perhaps, is that the population dispersal patterns claimed by this linguistic research is confirmed by archaeology and by DNA research. Taking this together with other work into the language roots of African and American Indian languages, Merritt Ruhlen of the University of Pennsylvania claims that an even more ancient 'mother-tongue' can be found. Called the 'Proto-world' language, it may be more than 40,000 years old. Perhaps the words in Genesis 11:6 were based on some truth: 'And the Lord said, Behold, the people *is* one, and they have all one language.' According to the Bible it was God who did '...confound their language, that they may not understand one another's speech. So the Lord scattered them abroad...' (Genesis 11: 7-8).

However people came to disperse, it seems likely from linguistic research that we once had a common language, while the wide dispersal of the Paleolithic goddess implies we had a common spiritual belief as well. As we have seen, Margaret Ehrenberg would disagree with this last point. This is her second reason why:

> The belief systems of forager and other small-scale societies, who are closely in touch with the natural world and whose own social systems are based on greater equality than that of later socially stratified societies, typically centre on general spirits and forces, rather than on personified gods and goddesses.[14]

But she fails to mention the worship of ancestors. People who today worship 'spirits', worship the spirits of natural things such as 'the forest', or spirits of their dead ancestors. These two concepts could have been combined in 'the Original Mother', the goddess, she who created all people, animals, plants and things, who resides in all natural things such as the earth, the forest, the sky, the water and, some might have believed, in animals and people as well.

Ehrenberg is also against the idea that some of the figurines represented priestesses for the goddess: 'religions involving deities, let alone priestesses with specialised functions, imply a political and social organisation far more complex than that likely to have existed at the periods in question'.[15] Yet even the most 'primitive' cultures have their spiritual intermediaries, usually referred to as 'shamans'. Human spirituality comes with the person and must, therefore, come before social or political organisation. Social organisation comes *later* and endeavours to incorporate or institutionalise spiritual thought. If, as Gimbutas and others assert, people once thought in terms of a parthenogenetic woman, and goddess, her intermediary would most likely have been a person closer to her 'in likeness' – a female shaman, a priestess.

The concept of a parthenogenetic goddess, however, is very ancient and widespread, and it may be our best clue as to what the Paleolithic figurines represent. The Egyptian goddess, Nun, who was associated with the primaeval ocean, created the sky-god and the rest of the universe, without a consort. Similarly – without a male consort – the Sumerian goddess Nammu created the sky-god and the earth. The Babylonian goddess Dunnu founded the line, as did Danuna in Crete, Danu in Anatolia, Danae in Greece and Diana of the Gauls. While in Ghana, Africa, the Akan goddess, Nyame, gave birth to the universe without a male consort, as did the Phoenician and Carthaginian goddess Tanit.

Spirituality is as old as people are; where you have people, you have spirituality. Today, people who have few material possessions will still have some symbol of their spiritual beliefs. A very poor Indian will have a shrine in his or her home. Urchins running shoe-less in the streets of South America have a crucifix around their neck, while the star of David provides the equivalent for Jews. These are the things we gather strength from. On spiritual symbols we focus our hopes, fears and devotion. Symbols are empowered, and empowering. They are our insurance and our escape. They are important and, probably, some of the first things we made. And what did we make? Symbols of women, women without men, the Creatrix – the parthenogenetic woman.

Ritual defloration

We know from many places and times that phallic-shaped objects have been used as tools for the defloration of virgins. The contexts in which this happened were many. Ritual defloration was a common practice in certain Indian priestly cults, a patriarchal context; while in certain African tribes, the context could have been matriarchal. On the subject, Nick Douglas writes:

> In India, and in certain tribes in Africa, girls were ceremonially deflowered with a symbolic representation of the Shiva Lingam, a ritual object kept especially for this purpose, or by squatting upon the Lingam of a statue of Shiva. Often this ceremony would take place in front of the whole tribe and would be an occasion of celebration. In some tribes, the defloration of virgins was performed by an older woman.[16]

Looking at some of the phallic-shaped objects of the ancient past one can't help wondering if they weren't used in the same way. On the next page is an illustration of a 23 cm-high goddess in serpentine marble, found in Italy and dated at 20,000 B.C., which is similar to many others in the shape of her body and pendulous breasts but, where the head would usually be, there is a penis, complete with an obvious vein on the underside.

We also have examples of what could be defloration objects from the very much later Cycladic culture, dated between 3300 and 2400 B.C. This culture thrived on the two hundred or so islands which circle the holy island of Delos in the Aegean sea and produced countless so-called 'canonical' figurines which are invariably the same despite the huge variation in their size from 10 cm to lifesize: marble, female, standing on tiptoe (or sometimes with no legs), with a head with a pinched nose, and arms which are folded right under left, across the stomach. Some are pregnant, but most are not. One figurine which could have been used as a ritual defloration object comes from the island of Amorgos: instead of a head it has a phallic protuberance and, very unusually, the arms are folded right *above* left.

If a person believes that women can produce children parthenogenetically, after the onset of menstruation, there is only one thing (possibly) standing in the way: the hymen. In the Trobriand creation myths the original mothers of the different islands have their hymens broken, in one case by falling rain, and in another by water dropping from a stalactite: 'The drops of water pierce her vagina, and thus deprive her of her virginity.'[17] This was discovered by Malinowski, whose conversations with the Trobrianders around 1916 led to the subject of the methods a woman might use:

In other myths of origin the means of piercing the hymen are not mentioned, but it is often explicitly stated that the ancestress was without a man, and could, therefore, have no sexual intercourse. When asked in so many words how it was that they bore children without a man, the natives would mention, more or less coarsely or jestingly, some means of perforation which they could easily have used, and it was clear that no more was necessary.[18]

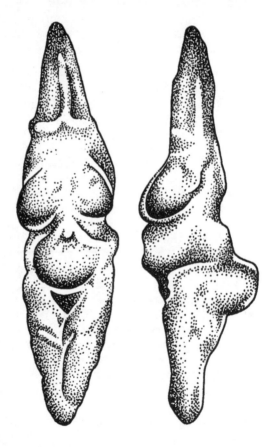

Savignano goddess
Carved in serpentine marble, standing 23 cm high. Found at Savignano in northern Italy. Provisionally dated around 20,000 B.C. although could be up to 7,000 years older. (Prähistorische Staatssammlung, Munich, Germany)

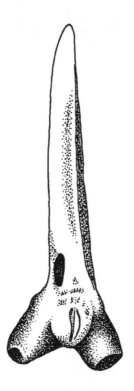

Le Placard vulva/phallus composite
Carved out of reindeer antler, this 15 cm high object depicts both
vulva and phallus. Dated between 15,000–13,000 B.C. it was found
at the cave of Le Placard, Charente, France.

I doubt that the Trobrianders would be shocked by the suggestion
that some of the phallic-shaped objects produced during the
Paleolithic and Neolithic eras were used in ritual defloration.
Even though they wouldn't use the objects themselves, they
could see the logic for them: a baby can't come out of a woman's
body unless the hymen has been broken. In the Trobriand Islands,
the hymen is broken during early sexual experience, of which there
is a great deal, and thereafter men keep the baby passage – the
vagina – open and lubricated so the baby can come out. His role
has nothing to do with the baby getting *in* – he contributes no
'seed', 'vital impulse' or anything else; it's about facilitating the
baby getting *out*. (Incidentally, everyone knows that virgins can't

have babies but the logic behind this knowledge says it's because her baby-passage is too tight and the ancestral spirits, the *baloma*, know it so don't enter her to be reborn.)

Allowing any potential baby an easy exit from the young woman's body might have been the logic behind ritual defloration, in certain cultures and at certain times. I don't think this need have been a terrifying experience. Indeed, I imagine gentle women eased the phallic-shaped object with oils, making the defloration easier perhaps than the usual method. But why use an object which has been carved into the shape of a female body, or which incorporates in its design women's reproductive parts – the vagina and breasts?

As we are talking about objects from historic periods ranging over 20,000 years, and from such a wide area, and made by people who undoubtedly had different things going on in their minds, there is no simple answer to this question. A phallic-shaped object might indicate that it relates to men but in the ritual defloration context, it is a *substitute* for the actual male part which, in the Trobriands for example, is seen as a tool for facilitating baby-exit, and a tool which could be replaced by other means, just as effective. The phallus, then, does not have to represent the delivery of male seed or even the stimulation of female seed (or her generativity in general). As it is an actual penis that usually goes into the vagina, it makes sense to make the ritual defloration object look like a penis. Why not? Sex is heaven on earth and one of the deity's greatest gifts to men and women alike, and available to any young girl who no longer has a hymen. The penis, as we know very well ourselves, has a role as pleasure provider, not necessarily as reproducer. And why not combine the male and female forms in one shape? – symbolic of the union of opposites, more powerful when in combination. Again, this doesn't have to have a reproductive connotation, rather a social and even metaphysical one.

Having a female shape attached to the phallic ritual defloration object could, perhaps, have simply signified the young girl's transition into potential motherhood; the female form of the phallic object representing the initiate's own new fecundity. Or perhaps the female form represented the goddess – she who facilitated all fecundity.

When ritual defloration objects are used in patriarchal contexts – representing the lingam of the god Shiva, for example – the object stands in for the original Creator, a male, and the society has in its material culture many other images of that god. In the Paleo-lithic, however, there is an absence of gods. They simply don't exist. Also, in patriarchal societies the most usual 'ritual defloration' involves the husband deflowering his new bride in a room or house which is surrounded by family and neighbours waiting to see the bloodied sheets hung out, evidence that the bride was a virgin, and the hymen was 'his' – the husband's. In this context, it is extremely important that the living male body enters the body of the woman, precisely because there is a chance of his seed being delivered into her. Clearly, when you use a ritual defloration object, the delivery of seed is not what it's about. Ritual defloration takes the hymen away from the husband and gives it to someone else. In a patriarchal context, it could symbolically be 'sacrificed' – the ultimate gift to the ultimate Creator. This may very well have been the object behind the activity in Paleolithic times, except that the ultimate creator was a Creatrix – a female. However, no 'sacrifice' may have been intended by the practice. Ritual defloration could simply have been the removal of the one barrier to the young woman's fecundity, a fecundity that came from within her and needed access to get out. Whether the hymen was removed by an actual penis or a ritual defloration object may have been immaterial in terms of actual reproduction, but crucial in terms of ritual honouring to the source of all creativity – the parthenogenetic goddess, mother of all parthenogenetic women.

14 Woman the prime mover

Precisely *because* women give birth, nourish from the breast and care for their children over many years, women became the prime movers in socialisation and civilisation. We learned to speak while in our mother's arms, not while out on the hunt. Mother taught us caring and sharing. She fed us, carried us around, protected us when we were in danger, and healed us when we were ill. Women developed social skills because they needed each other for communal child-care. The first tools were the digging sticks women used to gather enough food to feed the children, and the containers they devised to carry that food, and the slings they needed to carry the infants they held against their bodies as they worked. Women were the first horticulturists, and the first agriculturists; society developed around the spiritual places where the goddess was worshipped, and where women ruled supreme.

The cartoon picture of the prehistoric thug dragging a woman back to his cave by her long hair is a myth, as archaeologist Margaret Ehrenberg explains:

> the crucial steps in human development were predominantly inspired by females. These include economic and technological innovations, and the role of females as the social centre of groups. This contrasts sharply with the traditional picture of the male as protector and hunter, bringing food back to the pair-bonded female. That model treats masculine aggression as normal, assumes that long-term, one-to-one, male–female bonding was a primary development, with the male as the major food provider, and that male dominance was inherently linked to hunting skills. None of these patterns, however, accords with the behaviour of any but the traditional Western male. Other male primates do not follow this pattern, nor do non-Western human groups, in particular those foraging societies

whose lifestyle in many ways accords most closely with putative early human and Paleolithic cultural patterns.[1]

To construct a picture of the past we draw upon many scientific fields of enquiry, each one of which has a far deeper understanding of the facts than it had even a few decades ago. Primatologists study our nearest animal relatives, and palaeontologists work with geneticists to decipher the human fossil record, putting measurements through the computer so we can ascertain the relative body size between men and women, and see if there was any difference in the size of their canine teeth, which may indicate different ways of relating to one another. Historians examine and re-examine interpretations of the past, paleobotanists work with geologists to map out the landscape, while archaeologists dig. Meanwhile, all over the world, ethnographers record the habits, customs and differences among the many branches of the human race, with concepts of evolutionary biology ringing in their ears.

Evolution theory began, of course, with Charles Darwin. This is what he said in 1871: 'Man is more powerful in body and mind than woman, and in the savage state he keeps her in a far more abject state of bondage than does the male of any animal; therefore it is not surprising that he should have gained the power of selection.'[2]

Yet Darwin was aware that male animals have evolved all manner of devices to make themselves attractive to females and that, in fact, the female animal selects. He also knew of 'utterly barbarous tribes' in which 'the women have more power in choosing, rejecting, and tempting their lovers, or of afterwards changing their husbands, than might have been expected' – expected, that is, by the patriarchal Victorian male. These facts made Darwin's argument inconsistent, and he knew it. He could only suggest that in the ancient past women had more sexual initiative. A question mark was left hanging in the air.

From Darwin on, generations of male scientists and pseudo-scientists set about sorting out the evolution problem. It was suggested that women became pretty, delicate and sweet-voiced to assuage the male who had become a dominant thug. With very limited information, but plenty of sexist bias, we came to fill our libraries with nonsense.

Monkey talk

In the search for the 'natural' behaviour of humans, science takes account of the behaviour of our nearest animal relatives, the primates. But which monkey do you choose to extrapolate from: the baboon who lives on the savanna, or the baboon who lives in the jungle? They behave very differently.[3] Studies in DNA and proteins show that the African chimpanzee is a 'sibling species',[4] but as we diverged from a common ancestor at least five million years ago, we haven't now a great deal in common except our chromosomes. Nevertheless, as everyone else does it, let's talk monkeys.

My favourite monkey is the bonobo, a nearly extinct chimp living in the forests of Zaire which, in terms of chromosomes, may be our nearest animal relative. Although there is male and female pair-bonding, the bonobos are a remarkably amorous lot, indulging in a great deal of erotic activity, male–female, male–male, female–female and adult–child. The bonobos make love, not war. Aggression is very rare. Bonobo society is centred round caring for babies and the young, who take five years to wean. Scientists working on chimp intelligence choose the bonobos because they are smart enough to understand abstract symbols and language. They can even count. Were it not for the fact that they can't make consonant sounds, we might be able to have an intelligent conversation with them. They certainly understand what we say to them.

Working with a variety of primates, scientists have made the following findings. The central core of a primate group consists of several females and their offspring, including older 'boys'. In many primate societies 'this female core provides the principal social continuity of the group'.[5] The leading females decide when and where the troop move.[6] Shirley Strum reports that the females form friendships: 'I observed only two types of baboon friendships, those between females, and those between males and females – never between adult males.'[7] The female friendships were more intense and long-lasting. She writes: 'among baboons, as among most monkeys and apes, paternity is neither known nor recognised.'[8] In *Woman the Gatherer*, W.C. McGrew notes the behaviour of chimpanzees: 'females are solely responsible for childrearing, with no direct help from adult males'.[9] Males feed themselves; females feed themselves and the children. McGrew also says: 'In all cases that allow sex differences to be examined, it is females that use tools more commonly in feeding ... Female chimpanzees fish for termites over three times as frequently as males.'[10] However, 'male chimpanzees predominate in the use of weapons'.[11] They drop boughs from trees, charge, club, whip and throw. 'Male chimpanzees make virtually all kills of monkeys.'[12]

Although male primates fight among themselves, female primates stand up to males and form alliances against aggressive ones. Of the Japanese macaques, G. Gray Eaton writes: 'If a male fights with a female, other females will usually come to her aid, whereas adult males rarely assist one another.'[13] Of the langur, Phyllis Jay reports: 'Five instances were recorded of single adult females actually hitting or slapping a dominant adult male ... no male was ever seen to strike a female in return.'[14] While Jane Lancaster says of the vervet monkeys she studied: 'I often saw females form coalitions against the top three males when they monopolised a prized food source or frightened an infant.'[15]

It was once thought that a male's dominance within the group was entirely the result of his macho performance, but, as Zihlman reports, 'studies have shown the centrality of the mother rather than of males in determining the dominance status of offspring'.[16] Jane Lancaster observes: ' "Dominance" is more correlated with length of tenure in a social group than with winning fights.'[17] Females like males who share their food and groom; they oust aggressive males from the group. Moreover, 'dominant' males are not necessarily more successful reproductively.

In 1979 Caroline Tutin discovered that 50 per cent of impregnations among a group of chimpanzees had occurred while the female and her offspring accompanied a male in a 'consortship' – a private little party away from the main group. This was a surprising percentage in view of the fact that consortships accounted for only 2 per cent of observed matings. In consortship, the female has chosen her mate. The other 50 per cent of pregnancies occurred mostly during 'opportunistic' sex – when several males may copulate with a female. Fewer copulations occur in circumstances in which a male displays 'possessive' behaviour by trying to halt the amorous advances of the other males.[18] All this gives a very different impression to that perpetrated by the idea of a 'dominant' male monopolising his 'harem' of females and so ensuring his genes have a future. During 'opportunistic' matings the female has the chance to try out the various males, which helps her decide who she will have a consortship with later. Female primates initiate sex by drawing attention to their state of oestrus. A male might be 'dominant' but that just means he is the first in line because the other males are quietly waiting their turn. This whole pattern of reproduction means that 'dominance' does not equate with reproductive success; it is not therefore an evolutionary factor.

To summarise: male primates do not monopolise a particular female or group of females; there is no male-controlled 'family' or 'harem'; male primates do not initiate sex; male primates do not physically abuse females or their offspring; male primates do not lead the group. What male primates do do is hang around the females, trying to be impressive and nice, hoping to get laid.

Body and canine teeth dimorphism

The evidence is not yet clear as to whether there was any significant difference in body size between our most ancient male and female ancestors. Today, women are about 84 per cent the weight of men. The traditional (sexist) interpretation of this difference is that males became larger because the larger and, so it's assumed, more aggressive males won access to the females, thus ensuring their genes a place in the future. However, the difference in body size can be interpreted in quite another way. You could say that men became larger because, like male monkeys, they had nothing to do all day except feed themselves! Females have to gather enough calories to feed the children as well as themselves, consequently their share of calories is smaller and they don't grow so large.

There is, moreover, no evidence that early man was the aggressive thug he is so often portrayed as. Adrienne Zihlman explains the significance of the fact that men and women had canine teeth the same size: 'The fact that they are small and nondimorphic among early hominids could reflect greater sociability among individuals, between males and females, females and females, and a reduction of male–male competition and aggression similar to, and even an extension of, that found among pygmy chimpanzees.'[19]

'Pygmy' chimpanzee is another name for the 'bonobo' chimps – the ones who like to make love, not war.

Gathering

Before hunting there were probably about two million years of gathering. Women, of course, can gather as easily as men and with many mouths to feed, they have the incentive to do more of it. It's not simply a matter of picking fruits and nuts. Gatherers must know the difference between nutritional foods and poisonous ones. They must be able to recognise that under a few shrivelled leaves lying on the surface of the ground a tuber can be found. On a daily excursion a gatherer might find edible leaves, roots, barks, stems, bulbs, fungi and honey. They will pick up extra protein in the form of eggs, insects, lizards and other small animals. As fish bones do not endure the test of time we don't know for certain that our ancient ancestors went fishing, but it is likely. If they lived by the sea they might also have had the benefit of seaweed and molluscs.

Gatherers need to remember where food sources are, and they need to communicate the positive or negative value of different species to their children. It was thus much more likely to have been gathering that developed human language, rather than hunting which is, of necessity, a silent activity. Also, gatherers need containers to carry the booty home; and they need to devise equipment to hold an infant secure against their body while they work.

Adrienne Zihlman points out that 'behaviors attributed to hunting can as easily be explained by gathering: long distance walking, use of tools, sharing resources, large home range, home base, low population density, detailed knowledge of the environment, and cognitive mapping'. She speculates that 'Sharing patterns established between mothers and offspring continued into adulthood and expanded to include other adult males and non-related individuals.'[20]

It is often said that men invented tools when they started hunting. People came to this conclusion before they considered the fact that tools used in gathering are biodegradable. We thought stone tools were the first simply because time allowed them to be found; unlike digging sticks, they had not turned to dust! However, as Zihlman points out: 'As aspects of evolutionary continuity, wooden spears and, later, hafted tools, fishing nets, and the like may have developed from digging sticks or "bags" for collecting and carrying, in the same way that digging sticks may have emerged from an ape's termiting stick.'[21] Also, stone tools were probably used for many millennia before we started hunting. The Kung San women carry a stone 'chopper' to sharpen their digging sticks,[22] and Australian Aboriginals use organic tools which they make with stone tools.[23] Also, as Zihlman says, tools for hunting may have developed from tools used to cut fruits or pulverise vegetables.

It is also worth bearing in mind that the first evidence of tools used in connection with meat may not indicate hunting, but butchering. The famous Olduvai Gorge finds, dated between 1.6 and 1.8 million years old, were discovered in an area that was once a swamp. Mary Leakey suggested that the ancient hippo and elephants had become mired and, as Zihlman suggests, butchering such an animal 'could as easily have been done by women with children as by men'.[24]

Hunting

Ancient rock art from all over the world shows women armed with bows and arrows, as well as women taking part in hunting magic. As we begin to look at this subject from a non-sexist perspective it becomes clear that hunting is not a male preserve although, as time passed, it increasingly became one.

The idea that females are incapacitated because they have wombs is ridiculous when you consider the fact that the lioness not only hunts, but must hunt more than the lion because she, and only she, provides all the food for her young, for two or three years. And the idea that a woman is too small or weak to hunt is ludicrous when you consider that the Mbuti pygmies who live in the jungles of north-east Zaire in central Africa, standing at four feet two inches, can bring down elephant and buffalo. Obviously, size is not the key to hunting success. Indeed, the greatest handicap to women's participation in the hunt is other people's prejudice against the idea. Of the Agta negritos living in north-eastern Luzon, an island in the Philippines, anthropologists Agnes Estioko-Griffin and P. Bion Griffin write: 'It is no accident that only in remote mountain regions do women hunt. Farmers place considerable pressure on Agta to conform to lowland customs. All lowlanders regard hunting as solely a male activity and ridicule participation by women.'[25]

Left to their own devices, Agta women will hunt wild pig and boar, monkeys and deer. They go armed with a bow and arrows or, if they prefer, a machete and a pack of dogs. According to the Griffins: 'women not only hunt but appear to hunt frequently. Like men, some enjoy hunting more than others.' Biology is not much of a handicap:

Among the Agta, during late pregnancy and for the first few months of nursing, a woman will not hunt. In spite of the small size of each residential group, however, some females seem always to be around to hunt, although one or more may be temporarily withdrawn from the activity. Women with young children hunt less than teenagers and older women. On the occasion of brief hunts – part of one day – children are cared for by older siblings, by grandparents, and by other relatives. Occasionally a father will tend a child. Only infants are closely tied to mothers. Girls start hunting shortly after puberty. Before then they are gaining forest knowledge but are not strong. Boys are no different ... As long as strength to travel and to carry game is retained, people hunt. Our best informant, a young grandmother, hunts several times a week.[26]

As a matter of interest, aside from hunting, there are other ways in which Agta women do not conform to the stereotypical patriarchally-controlled female: 'Women are as vocal and as critical in reaching decisions as are men ... Agta women are actually more aggressive traders than are men, who do not like confrontation ... almost certainly girls are able to engage in sexual activity with relative ease'; and as far as extra-marital sexual relations are concerned, 'neither males nor females seem to be especially singled out for criticism – either sex may divorce the other with equal ease'. The Griffins conclude: 'Agta women are equal to men.'[27]

Another contemporary society in which women regularly hunt can be found in Zaire. Anthropologist Colin Turnbull tells us that because the whole community cares for its children, mothers are not incapacitated by their children:

Childbearing in no way diminishes the mother's importance in Mbuti economic life. Since they are perfectly capable of giving birth to a child while on the hunt, then rejoining the hunt the same morning, mothers see no reason why they should not continue to participate fully in all adult women's activities. For the first three days following childbirth only do they remain secluded in the dark shade of their hut, gradually introducing the infant to the light of the camp outside. They rarely stay away from the daily hunt longer than this, never in my records for more than a week. They may take the infant with them, slung at the side so that it has constant access to the breast, or they may leave it in the camp with one of the other mothers; in the larger camps there is nearly always one adult woman who stays behind for some reason, as well as some of the elders and children.[28]

Portrait of a 'primitive' society

Charles Darwin thought that 'in the savage state' men keep women 'in a far more abject stage of bondage than does the male of any other animal'. He probably got this impression from patriarchal cultures less genteel than Victorian England. Certainly, he hadn't been to the jungles of north-east Zaire and met the Mbuti. Luckily for us, however, anthropologists Colin Turnbull and Patrick Putnam have between them spent 50 years living with this tribe of pygmy hunter-gatherers and they can give us a clear, unprejudiced picture of their life. Turnbull writes:

> both men and women see themselves as equal in all respects except the supremely vital one that, whereas the woman can (and on occasion does) do almost everything the male does, she can do one thing no male can do: give birth to life ... the Mbuti associate womanhood with the life-giving principle and demonstrate this clearly in their rituals, formal and informal.[29]

The Mbuti home is the forest, 'which for them represents the supreme creative principle'. Sometimes they address it as *eba* (father) but 'most of the time they refer to it as *ema* (mother)'.

During his 25 years with the Mbuti, Turnbull lived the four stages of Mbuti life – as child, youth, adult and elder. As a child he was expected to address all women as *ema*, and 'discipline was exerted by any of my mothers'.[30]

Youths of both sexes 'indulge in premarital sex with enthusiasm and delight. They talk about it openly, with neither shame nor undue bravado'.[31] For girls, sexual life begins with the *elima*: 'When a girl first "sees the blood" and enters her *elima*, there is universal rejoicing' because 'now she can become a mother.' She can participate in the courtship which takes place around the *elima* house, and is a preparation for marriage:

Male youths from nearby hunting bands gather outside the
elima house every afternoon and wait for the girls to come
out. I later learned from the girls that they often deliber-
ately keep the boys waiting to annoy them and show their
power. They have an even more telling way of showing their
power, for to be invited into the *elima* hut, which of course
is what was on every boy's mind, one has to be chosen by
a girl, who issues the invitation by delivering a sound
whipping to the boy of her choice with a long, supple
sapling. Even that did not give us instant access. There were
the *elima* songs to be sung, a joint endeavour, in which the
boys had the subordinate role of providing a background
chorus. When the girls retired back into their hut, those
who had been whipped could follow, but now there was
another hazard. The moment the girls retired, the mothers
all gathered in front of the hut, armed with sticks and
stones, some with bows with which they fired small missiles.
No one entered unscathed and the mothers were perfectly
capable of preventing any undesirable from entering
whether he had been whipped or not ... Once inside you
were free to sleep with whoever was agreeable. As youths,
then, we learned that both girls and women have considerable power, and that while girls are exquisitely flirtable,
it is on their own terms.[32]

Pre-marital pregnancy is not a worry because the Mbuti believe
it can't happen if the partners 'hold each other by the shoulders
and not embrace fully'.[33] Naturally, Turnbull found this idea
intriguing and consulted with Putnam, but they couldn't record
one single case of pre-marital pregnancy nor could they find
'even a hint of abortion being the explanation'. Extra-marital intercourse is likewise not forbidden because the same 'holding
shoulders' position is used and so, according to the Mbuti, it cannot
result in pregnancy. When a couple are sure they're right for
each other, they build a marriage hut, move in, and start going
hunting together. The shoulder-holding during intercourse stops
and they '*know* that children *will* be born'.[34]

Married couples do not have sex for three years after the birth
of a child, during which the mother is breast-feeding.

... the mothers seem to sublimate their sexual drive through
their almost passionate concern for their infants ... I have
been unable to document any extramarital escapades on

the part of nursing mothers, not even flirtations. For their part, adult men and male youths seem singularly uninterested in such mothers as sexual partners. The men, however, both flirt and have sexual intercourse with other women.[35]

Elders are thought to be 'endowed with special spiritual powers because of their proximity to death and the "other world".' However, some seem to command more respect than others:

> (The female elder) has both authority and power. She may be a gentle, loving, and kindly old lady one moment, as many of the older women are, but in a flash she becomes pure power and is heeded by everyone. Ridicule is an important element in all conflict resolution; only the old women come out into the open, in the middle of the camp, and make explicit criticisms. Men may use the same central position, which commands attention, but only to grumble or complain and perhaps make minor and rather petty criticisms that are most likely to be ignored.[36]

The most important ritual object of the Mbuti is the *molimo* trumpet, 'symbol of the forest as spiritual entity'. Although the male youths usually have control of the *molimo*, that control is illusionary, as Turnbull discovered one night during a 'great *molimo*', a ceremony that accompanies important occasions, usually a death. The women intervened in the proceedings, seized the sacred banja sticks which beat out the special *molimo* rhythm and, with a length of twine, tied a noose around the neck of each male in turn so the men were all bound together. As he was entwined, the man stopped singing, a woman took over, and thus women took control of the song. The men had built a special fire, 'drawn from each family, female, hearth', but this too the women kicked asunder:

> The men throw the logs back together and, with a dance that imitates the act of copulation, fan it back to life. The women repeat their act of destruction, perhaps several times, making it more than clear that while the men may have the power, through the *molimo* trumpet, to invoke the beneficial spirit of the forest, the women have the ultimate power over life and death.[37]

Mbuti women are powerful and, contrary to what Darwin might have liked to think, in no 'abject stage of bondage'. Indeed, if anyone is in bondage, it appears to be the men! Neither are women incapacitated by their ability to bear children. Turnbull writes 'as a male I am aware that it is the males who feel, to some extent, incapacitated'.[38] However, Turnbull also tells us that there is no 'sense of superordination or subordination' and that the Mbuti 'work hard at emphasizing the complementarity of the sexes'. This 'work' involves a ritual tug-of-war in which no side wins, and the *ekokomea* – a transvestite dance in which each gender dresses as the other and ridicules each other mercilessly. On domination, Turnbull writes: 'If anything dominates, it is that prime quality of *interdependence*, in such sharp contrast to the *independence* our own society values so highly.'[39]

Clues to the past

As a clue to the past, contemporary hunter-gatherers are interesting because they prove the point that there is nothing innate about men controlling women. We see from them that women are not incapacitated by having children, and that the nature of women's work need not be limited by biology. Another important point is made by Frances Dahlberg: 'Women among foragers participate in religious ceremonies and rituals; and the gods, mythical founders, and shamans are both male and female. Witches may be of either sex; they are not predominantly male, as is common in pastoral or agricultural societies.'[40]

But, like us, contemporary hunter-gatherers have a past which, if it is not expressed in terms of monumental architecture, is told in terms of song, dance, ritual and ritualistic artefact. Quite often their history tells of a time when women had more power. The Mbuti say that women originally had the *molimo* trumpet but that men stole it from them. According to Catherine Berndt, there is a widespread mythical theme among Australian Aboriginal peoples that 'in the creative era women owned and controlled all or some of the most secret-sacred songs, rites, myths, and objects, to the exclusion or partial exclusion of men, but that men reversed this situation through trickery, theft, or persuasion'.[41] Similarly, among the Hadza of Tanzania, the ritual meat known as *epeme* used to belong to women and is now the property of men.[42] Similar stories of a male 'takeover' can be found in many people's creation myths and tribal fables and, if it's interesting to visit 'primitive' people, I think it would have been even more interesting, from a feminist perspective, to have visited them, say, 10,000 or 30,000 years ago.

Dahlberg says that contemporary hunter-gatherers display an 'interdependent cooperation between the sexes',[43] but what were relations like, I wonder, before co-operation? What if the Mbuti (and other hunter-gatherers) didn't 'work hard at emphasising the complementarity of the sexes'? Perhaps the authority women have because they 'can do one thing no male can do: give birth to life' would be emphasised. And what was it like to have been around at the beginning of human life, a pygmy in Africa between 100,000 and 200,000 years ago? Were we more like monkeys? Were women and their offspring, making up 75 per cent of the population, the cohesive, sharing core of the group, independently caring for themselves and their young? And if we then thought in terms of a creative principle, in spiritual terms, did the kudos that goes with creativity accrue to the female? Did men think themselves involved in procreation at all? Of course we shall never know.

Only one thing is certain: biologically, women have more to do with reproduction than men. By growing the baby in their womb, women daily feed and nurture the life inside. Whether that physical contribution is seen in terms of flowing blood, or placenta or umbilicus, women are physically giving of themselves in a way men cannot. And by actually giving birth, women are transformational in a way men are not.

These were the obvious physical facts when people started wondering where the spiritual aspect of life came into it all. What was the difference between life and death, and where did that difference come from? For human beings, as far back as we know, procreation is about spirituality as well as biology. We know life is magic, and we reach out and try to connect with it, our source. For the Mbuti, the 'supreme creative principle' is the forest, which they usually refer to as 'mother', *ema*. Indians on the plains of America used to sing to Thought Woman, she who thought them up. Nun, the Egyptian Goddess, created the universe without the help of a male. Paleolithic cave-dwellers hacked vagina shapes out of the walls and carved awesome, enigmatic, faceless female forms. These were the symbols which interfaced people with the creative spirit, connected them to it. It was through the feminine that life, both physical and spiritual, flowed.

15 Trobriands – islands in the sun

We can't go back in time and ask the people of early Neolithic Malta why they built their temples in the shape of women's bodies. Nor can we ask our Paleolithic ancestors what message the Venus of Laussel was meant to convey. And there is no contemporary culture which builds similar monumental architecture, or carves figures of women exclusively. All that has gone.

What we can do, however, is visit a place where the people don't think men play any part in procreation: the Trobriand Islands, in the south-western edge of the Pacific Ocean. They were made famous in 1929 when Bronislaw Malinowski published *The Sexual Life of Savages*. For reasons which will become obvious later, the Trobriands became known as 'the islands of love', but this was the passage in Malinowski's book which made me go down to the travel agent and book my ticket there:

> The idea that it is solely and exclusively the mother who builds up the child's body, the man in no way contributing to its formation, is the most important factor in the legal system of the Trobrianders. Their views on the process of procreation, coupled with certain mythological and animistic beliefs, affirm, without doubt or reserve, that the child is of the same substance as its mother, and that between the father and the child there is no bond of physical union whatsoever ...
>
> We find the Trobriands a matrilineal society, in which descent, kinship, and every social relationship are legally reckoned through the mother only, and in which women have a considerable share in tribal life, even to the taking of a leading part in economic, ceremonial, and magical activities – a fact which very deeply influences all the customs of erotic life as well as the institution of marriage.[1]

181

To get to the Trobriand Islands you have to go to Port Moresby, the capital of Papua New Guinea, and catch a flight going east. You fly for an hour and 20 minutes in a well-used Air Nuigini six-seater through the mountainous peaks of the mainland, disconcertingly half-covered in clouds, then over the Solomon Sea. In the shimmering blue of the water, an emerald isle set in gold appears and, with a loop, you land at Losuia Airport, a modest affair brimming over with good-natured Melanesian folk. You are driven down a dusty track to the only hostelry, the Kiriwina Lodge, which looks like something out of a Humphrey Bogart movie. Next door is the only shop. It sells cooking oil and cigarettes and very little else. This is not a place where you can buy postcards.

Although there are 14 islands in the Trobriand group, most of them are tiny. The largest island is Kiriwina, which is about 30 miles long and between two and ten miles wide. People live in scattered villages, in houses built on short stilts. The walls are made of woven leaves and the roofs are thatched grass. Villages communicate with each other by means of a drum – a dug-out log. It seemed very effective. People tend their gardens, their few animals and go fishing. They look happy.

According to the Trobriand life-view, men have nothing to do with the procreative process other than, with the penis, keeping the vagina open and lubricated so the baby can come out. The testes are thought to be only an 'ornamental appendage'. When Malinowski asked what they were for, 'a native aesthete' explained 'how ugly would a penis look without the testes'. The testes serve 'to make it look proper'.[2]

Reproduction is a joint endeavour between spirits and women, the spirits being the spirits of dead ancestors. The theory goes like this: a person's soul or spirit does not die. It goes to one of the furthermost islands, Tuma, where it leads a constantly rejuvenating, happy existence among other spirits. At some point, the spirit desires to return to the physical plane and, with the assistance of a controlling spirit, does so after leaping back in age and transforming itself into a small pre-formed infant. It then makes its way into the womb of a living woman, who is always from the same clan or sub-clan as the spirit-child. She will be the 'parent', as Malinowski documents:

> That the mother contributes everything to the new being to be born of her is taken for granted by the natives, and

forcibly expressed by them. 'The mother feeds the infant in her body. Then, when it comes out, she feeds it with her milk.' 'The mother makes the child out of her blood.' ... This attitude is also to be found embodied, in an even more telling manner in the rules governing descent, inheritance, succession in rank, chieftainship, hereditary offices, and magic – in every regulation, in fact, concerning transmission by kinship.[3]

The spirit-baby enters the woman's body in one of two ways. Either it enters her vagina as she bathes in the sea – the *baloma*, the spirits, having floated into the bays from Tuma on bits of flotsam or leaves; or it enters from the head, dripping down to the abdomen in the form of spirit-blood. Whichever mode of entry a particular Trobriand Islander believes in, and there are several versions of each theme, the process is not directly related to intercourse. Indeed, even a woman who has never been touched by a man can conceive if her vagina has been opened by some other means, as is the case with the mythical mother, Bolutukwa, who lost her virginity when the water dropping from a stalactite pierced her vagina as she slept beneath it.

As proof of their theory of reproduction, the Trobrianders offer two facts. First, unmarried girls who undoubtedly experience a great deal of intercourse do not usually have babies. Second, hideously ugly women, with whom no Trobriand man would dream of (or admit to) sleeping with, do become pregnant.

The first point raises the question of birth control or abortion. Birth control seemed unknown but Malinowski did hear about a magic which 'exists to bring about premature birth'.[4] It involves certain herbs. Such an abortion, as we might call it, involves no moral censure because 'To kill a spirit by black magic or accident is quite impossible; his end will always mean merely a new beginning'.[5] It is not death, just postponement or redirection because the spirit-baby has not arrived at an opportune moment. Marriage is the correct environment to bring a child into the world because then there is a 'father' on hand to help out.

In fact, a man is father in two senses: as we would know it, the man who stands in 'intimate relation' to the mother; and the maternal uncle. The father who lives in the house with the mother and children is *tama*, the man you and I would think of as father, the man who has a deep affection for the children, who cares for them, plays with them, and helps with schooling. But a man is also *kadagu*, father to his sister's children, who 'belong' to his village.

He has a certain authority over the children, and can ask for their help, as well as give help to them. It is his job to impart tribal custom, and he becomes more important to them as time goes on.

Women move patrilocally in marriage – to the man's village – but as children belong to the same clan as the mother, her children really 'belong' to her village and should know its particular customs. Wherever a child may grow up, his or her 'own' village is the one where the mother grew up and where the *kadagu*, the maternal uncle, still lives.

In the Trobriands, as elsewhere, children often resemble the man who 'stands in intimate relation to the mother', *tama*, but given the fact that there is no biological connection between *tama* and the child, how do the Trobrianders explain any resemblance? When Malinowski asked the question he got a 'stereotyped' answer:

> 'It coagulates the face of the child; for always he lies with her, they sit together.' The expression *kuli*, to coagulate, to mould, was used over and over again in the answers which I received ...
>
> One of my informants explained it to me more exactly, turning his open hands to me palm upwards: 'Put some soft mash (*sesa*) on it, and it will mould like the hand. In the same manner, the husband remains with the woman and the child is moulded.' Another man told me: 'Always we give food from our hand to the child to eat, we give fruit and dainties, we give betel nut. This makes the child as it is.'[6]

The first question I asked of the Trobrianders was, 'Have you heard about the ovum and sperm?' Since Malinowski, of course, they had – many times over. Had it changed their theory of reproduction? No, the ovum and sperm were irrelevant. This concept came as a shock to me. I asked more questions until I finally realised that there is only one important word in the Trobriander's reproductive vocabulary – *baloma*, spirit.

The details as to how an ancestral spirit reincarnates into a child were not important, which is why in such a small geographical area they casually accept many versions of the two themes of entry – by sea and vagina, or spirit-blood and head. Indeed, the islanders take very little interest in certain aspects of the spirits:

no one knows whose incarnation the infant is – who he was in his previous existence. There is no remembrance of past life in Tuma or on earth. Any questioning of the natives makes it obvious that the whole problem appears to them irrelevant and indeed uninteresting. The only recognized rule which guides these metamorphoses is that the continuity of clan and sub-clan is preserved throughout. There are no moral ideas of recompense or punishment embodied in their reincarnation theory, no customs or ceremonies associated with it or bearing witness to it.[7]

I found the Trobrianders' lack of interest in the physical facts of life puzzling, but then I come from a culture that has to know everything. If we thought we were reincarnated from ancestors, I'm sure there'd be an even more flourishing trade in regression therapy than there already is, where we'd find out who we'd been before, for many generations up the line. But the Trobrianders are altogether more casual about where they came from. And the ovum and sperm didn't interest them at all. They were well aware of the scientifically proven facts, but they just didn't care. The physical details are irrelevant. So the man puts something, so the woman puts something. So what? The body still grows in the womb, and the child is still a *baloma*, transformed.

Yet another way of life

The Trobriand life-view weaves a complex web throughout the whole of life, and death. Nobody dies, of course. They just go to Tuma. Because children are actually one's ancestors, usually a maternal relative, parents don't boss their children around. Because the mother is the 'parent', children go with her in the case of divorce. Because sex has nothing to do with procreation, it is very free.

All this came as a shock to the western world in 1929 and there were those who said Malinowski must be mad. There still are. But I couldn't find anybody in the Trobriands who thought Malinowski had got it wrong, and judging from what I saw, I'd say that, considering he did his field-work between 1914 and 1918, he probably had it right. Obviously things have changed since then, but not so much you don't recognise you're in a different world.

It goes without saying that there is no preference for baby boys, and the idea of infanticide is abhorrent. Children are not property to be ordered around by adults. Malinowski never heard a parent giving a child a command – do this or do that – and the idea of telling a child off, or punishing it, was not only 'foreign' but 'distinctly repugnant'. From the age of four or five the children join the children's republic – a 'community within a community'. Malinowski comments: 'If the children make up their minds to do a certain thing, to go for a day's expedition, for instance, the grown-ups and even the chief himself, as I often observed, will not be able to stop them.'[8]

Sex starts at an early age:

> If we place the beginning of real sexual life at the age of six to eight in the case of girls, and ten to twelve in the case of boys, we shall probably not be erring very greatly in either direction ...
>
> As they are untrammelled by the authority of their elders and unrestrained by any moral code, except that of specific tribal taboo, there is nothing but their degree of curiosity, of ripeness, and of 'temperament' or sensuality, to determine how much or how little they shall indulge in sexual pastimes.[9]

There are no derogatory terms like slut or nymphomaniac in the Trobriand Islands. If you say a girl is a virgin, that's an insult. I was told that the girl with the most boyfriends on the go at the same time is considered the belle of the village. Pre-marital lovemaking is done in the bushes or bachelors' house.

Marriage is a pretty casual business. All you have to do is sleep together regularly, or be seen spending time together in public. Once it's established by the community that you're an item, the elaborate gift-giving system which is employed in all important aspects of Trobriand life comes into play. It's not unusual for a person to be married three or four times in a lifetime. Divorce is simple and, again, involves the giving of a gift; in the case of a subsequent marriage, the current husband gives a gift to the former, compensating him for the gift he gave the woman's parents.

Adultery is condemned, whether carried out by the woman or the man, only because it is emotionally upsetting for the spouse. However, according to Malinowski, if a woman has a child in the husband's absence he will not assume she has been adulterous and will cheerfully accept her child like any other. If a woman isn't happy at home she simply leaves and, naturally, takes the children with her. She moves her belongings to her mother's, or to a close maternal relative's and, if she wants, enjoys full sexual freedom. Invariably, the husband tries to make it up by sending peace offerings.

Malinowski was struck by the friendly and equal nature of married life, noting that men were helpful with the children and domestic chores. The women joined in freely with conversation and jokes, working independently, 'not with the air of a slave or a servant, but as one who manages her own department'.[10] If something needs doing, the wife asks the husband to do it. Although the husband is 'master' of the house in that the house belongs to him and is in his village, the wife is, after her brother, the legal head of the family. The wife has her own possessions and she and her family make a major contribution to the food supply of the household. In all, her status is equal to his or, to put it another way, his status is equal to hers.

The Trobriand Islands have a reputation as the quintessential Pacific islands of love, and I can understand why. One day I was standing talking to a 17-year-old boy when he raised his hand and started manipulating my nipple, all the while continuing the conversation and looking me straight in the eyes. I'd never seen such a bold come-on in all my life. What's more, I was having twinges in my uterus. This lad knew what he was doing, and he didn't learn it in a book. Another young man told me how he'd had sex with a girl, they both orgasmed twice, then she asked if he would do it again. He wasn't up to it; they parted. Some time later, he met her on the road and asked if they could have sex again. She laughed. With you, who can only do it twice? No way, she scoffed.

You might think the Trobrianders were just living up to their reputation, and having me on, but if there is a huge conspiracy-joke going on, it has involved every foreigner I know of who's been there. It is difficult to convey the totally different attitude these people have towards sex; it is disconcerting. One man I met there, a monied internationalist involved in a wholesale catering trade that employed many Trobriand women, told me that, educated and sophisticated though he was, he was reduced to a shy little boy by the frank, to say the least, questions of his workers and their piercing, direct gaze.

I wasn't in the Trobriands during the annual yam-harvesting festival, the highlight of the sexual calendar, but apparently everyone meets at night and dances to the hypnotic beat of drums. The dance is basically a grind – crotch to crotch with the thighs fanning outward, alternately with each step, accentuating the grind. The women wear short grass skirts and the men, the flimsiest of sarongs. The established, deliberate beat of the drums, so I am told, soon arouses a distinct erotic urge in the participants. Either gender can invite someone to dance, and lead them off the dance area and into the bushes for lovemaking. This is an opportunity for married couples to indulge their fancy because, ideally, during the rest of the year they should be faithful.

Now you know why missionaries set up residence in the Trobriand Islands! They've tried hard to change things, but the old ways show through. Lucky for us the first foreigner to invade the place was Malinowski, an anthropologist concerned only to record accurately what he saw. No archaeologist working in the distant future could have guessed from looking at the stilt-holes and rusty Coca-Cola tins that here was a place where reproduction had little or nothing to do with men. How many other places like this have been lost to posterity, I wonder?

Conclusion
Where to now?

Patriarchy was inevitable. The male-seed idea came in around 3000 B.C. and slowly spread until by 500 B.C. it had taken hold in the minds of most people we know about. Pockets of resistance lasted for a further thousand years. From these times, until the beginning of the twentieth century, when my grandmother was born, people were operating with a set of facts of life which led them directly to other understandings. They seriously believed that children only come from their fathers, so fathers were given authority over them. Ultimately, the source of the male 'seed' would have to be a male God. Because men had apparently been given exclusive rights of procreation, they were thought essentially different to women, more chosen, more creative and spiritual. Because biological inheritance was thought to go through the male line, inheritance of land went that way too. Girls were an evolutionary dead-end. Boys, on the other hand, were the hope for the future. Around these understandings society was arranged.

It was thought that women did not themselves reproduce, yet their bodies were designed to have babies. It seemed as if women had been put here on earth to do certain jobs for men: to have their babies and to feed them. Women were defined as men's 'helpmates', and so defined by 'nature', that is God, and to challenge this role thus became blasphemous. For women, it was a difficult situation. They were baby-making machines, but they had no rights over the babies, so if they left the man they would have to leave the children behind. The mothering instinct had mothers trapped.

Because men thought children were either 100 per cent their grown seed, or 100 per cent the grown seed of some other man, they became very touchy on the subject of paternity. Women became prisoners of men's paranoia. Although men were

189

considered the sole source of reproductive material, only women had the means to bring it to fruition. Their severe lack of reproductive control led men to control the baby-making machine. Women didn't argue about it because they too thought they had been designed by the Creator to nurture *men's* babies. In some important sense, they were not their own person; they were an appendage to men.

Men thought they had a right to control women. Indeed, to respect their ancestors and make sure their inheritance went down the right seed-line, men thought they had a *duty* to control women's sexuality. To make sure they had no access to sexual temptation, women became prisoners in a man's home, father's or husband's, and denied access to the outside world. Because she wasn't given the 'spark' of life, so to speak, from God, a brilliant woman was thought of as a sort of intellectual transvestite, a bit out of place. Men were the 'sentient soul', as Aristotle put it, the thinking, creative part, and millions of other men, and women, thought it too. Women had no future, no road into eternity through their genes, or seed or whatever, so the future was not theirs to have a say in. That's why women couldn't get any power. It reflected the fact they had no reproductive power. The political and financial future was none of their business.

All these many dynamics became our culture, our tradition. We now have a view of the 'nature' of men and women that is a synthesis of thousands of years of thinking that was, itself, incorrect because it was based on inaccurate information about the 'nature' of the reproductive relationship between men and women. Do we now in fact even know what the nature of men and women truly is? Can we see beyond the patriarchal veil of illusion that history has put between us and our true selves? As we enter the new reproductive theory era, of equal genetic input, we face a blank sheet upon which we shall have to write our own facts of life, our own definition of the 'nature' of men and women, and plan the future of the male–female relationship.

There is no point in blaming men for having the facts of life wrong. The male-seed view was logical enough, given the information they had to hand at the time. And it was only fair, women had had a monopoly, or near-monopoly on reproduction for a very long time. The balance was due to be redressed. The pity of it was, it necessitated men controlling the means of production, woman. She lost control of her life. Historically, it also introduced the whole concept of one human being controlling another, which had not been necessary when woman was thought to be both the source of reproductive material and the means of production. When 'mother' was the reproductive gender, she didn't *need* to reach outside herself and control another human being, but to ensure his reproduction 'father' did.

The male-seed reproduction theory was the mortar which kept the towers of patriarchy together. It held the stones of inequity in place. But the theory is now out of date so the mortar will crumble, and the patriarchal towers will inevitably fall. They are already falling around me here in London, although the House of Lords still cannot bring itself to pass a bill Lord Diamond has been trying to get through the house, allowing the peerage 'to descend to the eldest child of the body, lawfully begotten, whether male or female'.[1] It seems a fair-enough request but, in some places, boys are still preferred. Tradition lingers all around us, like a stale aroma in the air.

Elsewhere, the patriarchal tradition has a steely grip on the destiny of women. Today, there are in the world an estimated 100 million women who have been through some form of genital mutilation, and an additional two million young girls are at risk each year.[2] This cruel and dangerous practice is a testament to the all-male-seed paternity paranoia, coupled with women's ignorance of their reproductive rights; the former, driving men's demands for chastity, and the latter giving women little say in the matter. Because men had total proprietorial rights over children, we allowed the tradition of abuse. Today throughout the world there are an estimated 100 million young girls being raped by a close male relative, usually the father.[3] Not all mothers are ignorant of what is going on, but they turn a blind eye because to some degree they feel the child is his to do with as he wants – an apple from his tree. A friend of mine comes from a remote village in Britain where she once lived with her parents and six siblings. The father beat the boys and sexually abused the four girls and although everyone in the small community could guess what was going on, nobody did a thing. My friend says they took the attitude 'his house is his castle', and he can do what he wants. This was just 15 years ago, and here and around the world, it carries on.

Men thinking they have the right to control women results in men beating women, marital rape, rape by acquaintances and rape by strangers. Indeed, male control has taken us to the point where women are out-and-out terrified of men. The women I know are terrified that a man might break into the house at night and rape or murder them. Or that a man might rape them on the street. Many a woman has been surprised by a violent attack from her own man, let alone a stranger. Mothers are terrified of the men who might abduct their child, and even terrified of the man within the house who may abuse them too. Men say we need protecting. Yes, we need protecting from *them*.

There are wars in every corner of the earth, and on our own shores and streets. And we cannot call human development a great accomplishment when each and every day 35,000 children die of preventable disease or hunger.[4] This fact, and the endless wars, are an affront to humanity, not a blueprint for it. Men say they fight wars to protect us – the women and children. This is hogwash. Men fight wars because they can't think of anything more useful to do, like reading to the children or cooking dinner. They've been allowed to lounge around and get restless. Perhaps, as in earlier matriarchal times, men should stay home and weave and let women do the politics and arrange the resources. I don't think we would be so violent or so heartless.

I get angry when men say 'What is my role in life?' What, there isn't enough to do? But being *helpful* isn't part of the male role (that is supposedly women's role). *Control* has become men's thing. Take away their control of women and men have an identity crisis, wandering around lost, not knowing where or who they are. Take away their control of other peoples in the arena of war, and they feel they've lost their role in life. But there is plenty of scope for heroism in this world, the heroism of peace and life not of war and death. If all the gun-toting men in this world became Greenpeace heroes, we might just survive, and our grandchildren might just have a planet to grow up on.

Men have sunk into a state of moral turpitude, made possible by an inflated sense of their own importance and concomitant lack of understanding about human equality, plus the idea that they have a right to control women, children and anyone else they can lay their hands on. It is usually men who rape, rob and murder. It is usually men who spend their nation's wealth on armaments they use to oppress their own people. It is usually men who build nuclear weapons they plan to unleash on innocent people. And it is usually men who, in positions of power across the world, demand bribes before they will do the smallest thing. Men are in control, and totally *out* of control.

Many people think violence is men's 'natural' state. People tell me the world will never change because men are just thugs and women just too weak to oppose them. This idea has been around for a long time. The man who first stood up in the House of Commons and spoke in favour of voting rights for women, John Stuart Mill, said in *The Subjugation of Women* in 1869 that 'the inequality of rights between men and women has no source other than the law of the strongest'. If this were true there'd be little hope for the future, but Mill also says woman's 'state of bondage to some man' was caused, in addition to her 'inferiority in muscular strength', to 'the value attached to her by man'.[5] And here he has it. Women were given low value, as a consequence of their low reproductive value, and men took advantage.

It will take a lot of education for things to be put right. Men need to understand *why* they feel superior, *why* they have been chosen over the years to take leadership roles in politics, business and theology. It's not because they are better at it or more suited to it; it's because women weren't given a chance. *They* were defined as mere helpmates. And so, today, you find women doing the laundry at eleven o'clock at night, while the man sits around watching the news on TV. Men *still* think women are their naturally defined helpmates, born to wash, clean, shop and cook for them, and take care of the kids.

Not all men are like this, of course. Indeed, it's the good ones that remind you how bad the bad ones are. Man, by definition of the fact that he has testosterone, does not have to be a thug or even a bully. And men in Sweden and Holland prove that men don't have to be lazy around the house. They even wash the toilets, and don't think themselves any the less a man for it. In fact, they'd be affronted if you suggested they were so incapable they couldn't take care of themselves.

In the discussion about the ancient past, pre-history, you rarely hear about sexual ecstasy, but that is possibly our strongest evolutionary force. Women did not evolve to be controlled by men, with their clitorises already cut off; they are born with a body designed to experience intense sexual pleasure with a man. Likewise, men had every incentive to stay with women. They made them feel good. So there we are, enjoying each other; who needs control? Only the men who can't get sex and force themselves on women, and they can be thrown out – by a group of women if necessary. I can't see the other men standing up for them, why should they? They might lose their sexual access and might lose out when the products of women's gathering are being shared out.

In the long history of humanity, where children fitted into the picture has been and continues to be a matter of opinion. Today, there are very many people in the world who think children come entirely from the father, a very few who say it is from the mother alone, and people like us who say it comes from both. No doubt societies can be arranged to accommodate any view. People just get on with it, and take from the situation what it has to offer. Millions of women living with the male-seed theory today are quite happy taking care of their husband's children, while a Trobriand man takes care of his wife's children, according to Malinowski, being a 'hard-working and conscientious nurse'. He writes, 'the father is always interested in the children, sometimes passionately so, and performs all his duties eagerly and fondly'.[6] These men don't think they're the father but they still enjoy the child, and are involved with it socially, also having other paternal duties to their sister's children. They are not 'fathers' as we understand it, but they are very fatherly.

Technically, it is not difficult to come to the conclusion that there is equal genetic or 'seed' input. You could just say the man's white part, the semen, makes the white parts of the baby's body like the bones and nerves, while the woman contributes the red part, the menses, which makes the flesh. And other theories of equality could be arranged with the available facts, if so desired. During the 'interface years', when people knew both men and women were involved in reproduction, but weren't quite sure how, they may have agreed on a point of perfect reproductive balance, and lived in perfect harmony, for a few years at least. Equality in terms of reproduction could have been achieved in a thousand places over the years, and may still exist in remote corners of the world. We actually know very little about other culture's reproduction theories because the questions ethnographers ask don't go deep enough. Most of all we need to know if positive or negative values are attributed to the male and female contributions, and whether decisions have been made that one is 'best'.

Aristotle and a thousand eminent gentlemen of history have looked at the facts of life, as known, and drawn philosophical conclusions from them. We could do the same. We could say the sperm is minuscule compared to the ovum, so women are more important than men. We could say the ovum actually replicates, and deduce men have little part in the real 'matter' of the child. Because the sperm drops its packages of mitochondria in the tail, as the head alone enters the ovum, all human life gets these little generators of power from the mother. So you could say women are the powerhouse behind humanity. Also, because female sperm are slightly larger and heavier than male sperm, and come with 3 per cent more DNA, we could say girls are better than boys.

But extrapolations are a waste of time. The fact is that when the ovum and sperm meet, they metamorphose into something else. It takes about 20 hours after the sperm meets the ovum for the one-cell ovum to become a two-cell conceptus. And in those 20 hours a great number of exciting processes take place, including rotation, electrical activity, and the movement of chromosomes. From a chaotic collection of cells, order emerges. When it counts, at reproduction itself, the ovum and sperm are only so defined for 20 hours at very most, then they cease to be. Something else comes in their place, a new life. Are we to extrapolate from the ovum and sperm and reach philosophical positions, and arrange our lives around them, for the sake of 20 hours or less? The fusion of ovum and sperm is the miracle of reproduction, not one bit or the other. It is mirrored, of course, in the fusion of sexual ecstasy, that moment of bliss when we realise our oneness with the universe, the irrelevance of physical divisions, and the illusion that is gender and much else besides.

There is nothing *biological* standing in the way of equality for men and women. Children don't hold women back, men do by not taking equal responsibility for their children and home. They run the country but won't arrange for child-care. A woman is not the sum total of her hormones, any more than a man is. One body-builder in the gym may be an equalist, and his buddy not. They both have testosterone, but they think different things, and treat their women differently. That's why there is hope for the future. We can all change our minds.

From a reproductive point of view, we may actually be facing a serious danger. As far as can be said from the limited data available on the subject, sperm counts have reduced over the past 50 years by 50 per cent. This is according to figures compiled in Sweden, Britain and the USA. (There are no old records of sperm counts from the developing countries, but it is interesting that recent data show that men there have the same count as men in the West.) Another worrying feature is that sperm samples include high numbers of sperm which are incapacitated in some way, by having a deformed head, or no tail, two tails, being all tail, having a neck enlarged with cytoplasm, being hyperactive or having no movement. Sperm, being the smallest body cells, are vulnerable, and it's thought they may be suffering because of pollutants carrying man-made oestrogens, particularly in the water. Other reproductive problems are also rising sharply in incidence, such as testicular cancer which has risen 300–400 per cent in some places, while the number of boys with undescended testicles has doubled in the last 30 years.[7] All these are worrying developments for those concerned with our reproductive future, whether they are caused by pollutants in the water, or by ozone depletion, or some other cause. The trouble is, we haven't had sperm research for very long, and to know how our children are being affected, we may have to wait 20 years. It will take that long to get a perspective on the subject, because the old figures may be an aberration, and what we have now is normal. Only time will tell. In the meantime, we need to keep a sharp eye on the sperm counts. If they really are reducing so dramatically, we may be heading for serious reproductive trouble. Already, we're not doing well with one in six couples being unable to conceive. About a third of the time infertility is caused by difficulty with the sperm, a third of the time it is a problem with the woman's body, and in a third of cases, it is a combination of factors from both sides. At present, the process of sperm production is still a mystery, and science can offer little help to men with low sperm counts and other sperm problems.

If sperm do succumb to environmental pressures, humanity will have to fall back on the ovum. As the largest cell in the body, it seems to be more tough. Also, it is reproductively extremely capable, and it is not outside the realms of possibility that if science has to put its resources into finding a way to make the ovum create life on its own, the parthenogenetic woman may be with us. Women will give birth to clones of themselves, and there won't be a man in sight.

In the meantime, we're genetically engineering plants, animals and people. Pigs are bred using artificial DNA which allows their hearts to be used for human transplant; while many lives have already been saved by the sheep who are genetically engineered to produce human hormones. The divisions between humans and animals are becoming blurred, perhaps not surprisingly considering the difference between chimp and human DNA is a mere 1.6 per cent. Even plants are being genetically engineered to produce plastic, so does that make it a vegetable or a mineral? And where will it all end? I wouldn't worry about a future of impressive

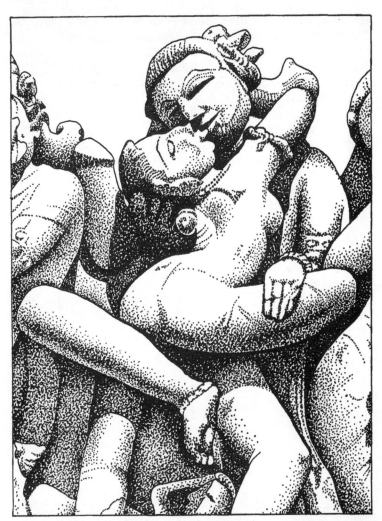

Temple sculpture, Khajuraho
Detail from the Tantric Vishwa Nath Temple (south wall) at Khajuraho,
India, built in sandstone during the 11th century A.D. Tantra, which
may be the oldest continuous philosophical system known by human
beings – and has been much repressed over time – holds that human
sexual libido is identical with the creative energy of the universe.

technological feats, if the world was replete with sensible people. But it is not. The potential for mismanagement of reproductive technology in the future is mind-boggling, whether considering human reproduction, animal husbandry, animal–human co-productions or plant-breeding techniques (not to mention bacteria and viruses). And things are moving along apace. By 2003 we may have before us the whole human blueprint, the decoded DNA. We already know maleness is conferred by the SRY gene (sex-determining region Y gene), and can turn female mice into males by injecting it into the fertilised mouse egg.[8] No doubt this news will be welcome on the Indian sub-continent – soon they can have all the boys they want.

It is important that we get our priorities right before going too far along the road of technological advance, and in terms of priorities we clearly still have a long way to go. Many difficult ethical decisions will face us all in the future and I'd feel a lot happier if there were more than a few women in the decision-making processes, both here and abroad. If we are going to face important matters of reproduction, in terms of humans, animals or plants, it would be better if we faced them together. That way we might get a balanced view.

Notes

Introduction

1. Aeschylus, *Plays Two. Oresteia: Agamemnon, Libation-Bearers, Eumenides,* p. 114.
2. John Mason Goode, 'Generation' in *Pantologia, A New Cyclopaedia* (London: Kearsley Walker et al., 1813), Col. 5.
3. Baer, *Autobiography of Dr. Karl Ernst von Baer,* p. 218.
4. Carr, *The Sexes,* p. 14.
5. Johnson and Everitt, *Essential Reproduction,* p. 222.
6. Delaney, *The Seed and the Soil,* p. 8.
7. Ibid., p. 53.

1   Man the seed, woman the incubator

1. Delaney, *The Seed and the Soil,* pp. 32–3.
2. Ibid., pp. 55–6.
3. As *erythrinus* means red, it's thought Aristotle meant *serranus anthius,* a fish that is brilliant red.
4. Aristotle, *The Generation of Animals,* II, I, 732.a. 3–10.
5. Cole, *Early Theories of Sexual Generation,* pp. 207–8.
6. R. Bradley, *A Philosophical Account of the Works of Nature* (London, 1721).
7. Erasmus Darwin, *Zoonomia; or the Laws of Organic Life,* Vol. 1, (London: Walter Scott 1794), pp. 482–537.
8. Horowitz, 'Aristotle and Women', p. 197.

2   How biology became destiny

1. Aristotle, *The Generation of Animals,* II, I, 727.a. 5–10.
2. Ibid., 727.a. 27–30.
3. Ibid., 737.a. 27–28.
4. Ibid., 775.a. 13–21.
5. Ibid., 732.a. 3–10.
6. Nancy Tuana, 'The Weaker Seed – The Sexist Bias of Reproductive Theory', in Tuana (ed.), *Feminism and Science,* p. 153.

7.   Aristotle, *Politics*, 1, 5, 1254b 12–16.
8.   See Basim Musallam, *Sex and Society in Islam* (Cambridge University Press, 1983).
9.   Delaney, *The Seed and the Soil*, p. 154.
10. Albertus Magnus, *On the Inner Secrets of Women*, quoted in Morgan, *A Misogynist's Source Book*, p. 221.
11. Galen, *On the Usefulness of the Parts of the Body*, 14.II.295.
12. Ibid., 14.II.301.
13. Delaney, *The Seed and the Soil*, p. 47.
14. *Autobiography of Dr. Karl Ernst von Baer*, p. 217.
15. Ibid., p. 225.
16. Geddes and Thomson, *The Evolution of Sex*, pp. 270–71.
17. Ibid., p. 267.
18. Galen, *On The Usefulness of the Parts of the Body*, 14.II.297.
19. Ibid., 14.II.299.
20. William Keith Brooks, *The Law of Heredity* (Baltimore: John Murphy, 1883).
21. Ibid., p. 257.
22. Judith Genova, 'Women and the Mismeasure of Thought', in N. Tuana (ed.), *Feminism and Science*, p. 214.
23. Marcel Kinsbourne, 'If Sex Differences in Brain Lateralization Exist, They Have Yet to be Discovered', *Behavioral and Brain Sciences*, 1980, 3: pp. 241–2.
24. Nancy Tuana, 'The Sexist Bias of Reproductive Theory', in N. Tuana (ed.), *Feminism and Science*, p. 156.
25. Ibid., p. 168.
26. Erasmus Darwin, *Zoonomia; or the Laws of Organic Life*, Vol. 1 (London: Walter Scott, 1794).

3   A resemblance to mother

1.   Cole, *Early Theories of Sexual Generation*, p. 85.
2.   Ibid., p. 73.
3.   J. Astruc, *A Treatise on all the Diseases Incident to Women* (London, 1743), quoted in Cole, ibid., p. 84.
4.   Cole, ibid.
5.   See Needham, *A History of Embryology*, p. 29.
6.   Cole, *Early Theories of Sexual Generation*, p. 60.

4   The word of God

1.   Raghib, *Safinat,* ed. Bulaq (Cairo, 1286h.), p. 15ff.
2.   Delaney, *The Seed and the Soil*, p. 156.
3.   Ibid., pp. 33–4.
4.   Ibid., p. 60.

5. Bouhdiba, *Sexuality in Islam*, p. 11.
6. Delaney, *The Seed and the Soil*, p. 288.
7. Monica Furlong, *Independent*, 15 December 1986.
8. Ginzberg, *Legends of the Bible*, pp.108–9.
9 Delaney, *The Seed and the Soil*, p. 231.

5   The ramifications

1. Delaney, *The Seed and the Soil*, p. 36.
2. Ibid., p. 161.
3. Gregersen, *Sexual Practices*, p. 276.
4. *Assignment*, BBC TV, 28 September 1993, producer Gizelle Portenier, reporter Emily Buchanan. This programme won the Gold Nymph award for best documentary at the Monte Carlo television festival, 11 February 1994. (See also Tahir Shah, 'Scanning for Boys', *Marie Claire*, November 1993, p. 81.)
5. Personal communication with Gizelle Portenier.
6. Jonathan Mirsky, 'China's baby girls "killed by the million"', *Observer* 26 January 1992, p. 13.
7. Miles, *The Women's History of the World*, p. 117.
8. Horowitz, 'Aristotle and Women', pp. 197–98 (Note 45).
9. Ibid., p. 198.
10. Boswell, 1776, Vol 2, p. 428
11. Delaney, *The Seed and the Soil*, p. 167.
12. Miles, *The Women's History of the World*, p. 112.
13. Debbie Taylor/New Internationalist Collective, *Women, a World Report* (London: Methuen, 1985), p. 65.
14. John Stuart Mill, *The Subjugation of Women*, 1869.
15. Quoted in Morgan, *A Misogynist's Sourcebook*, p. 185.
16. Quoted in Miles, *The Women's History of the World*, p. 103.
17. Webb, *Pathologica Indica*, pp. 259–78.
18. Malika Belghiti, 'Les Relations feminines et le Statut de la Femme dans la famille rurale', *Collection du Bulletin Economique et Social du Maroc* (Rabat), 1970, p. 24.
19. Ulla-Britt Engelbrektsson, *The Force of Tradition: Turkish Migrants at Home and Abroad* (Goteborg: Acta Universitatis Gotoburgensis, 1978).
20. Ginzberg, *Legends of the Bible*, p. 49.
21. Ibid., p. 36.
22. Cellmark Diagnostics (division of Zeneca plc), Unit 8, Blacklands Way, Abingdon Business Park, Abingdon, Oxfordshire OX14 1DY, tel: 0235 528609. The cost is £145 + VAT per each sample of blood.
23. El Saadawi, *The Hidden Face of Eve*, p. 41.
24. Miles, *The Women's History of the World*, p. 270.
25. El Saadawi, *The Hidden Face of Eve*, p. 201.
26. Mernissi, *Beyond the Veil*, p. 111.

27. *Under the Sun: The Women Who Smile*, BBC TV, 14 June 1990, producer Joanna Head (Bristol).
28. M. Barrett and S. McIntosh, *The Anti-Social Family*, (London: Verso, 1982).
29. *Criminal Statistics Digest 2*, ed. Gordon Barclay, (London: Home Office Research and Statistics Dept, 1993). (Figures are for 1991, from totals of 297 female and 378 male homicide victims.)
30. Gregersen, *Sexual Practices*.
31. Minces, *The House of Obedience*, p. 63.
32. Al-Bayhani, quoted in Yvonne Jazbeck Haddad, 'Traditional Affirmations Concerning the Role of Women as found in Contemporary Arab Islamic Literature', *Women in Contemporary Muslim Societies* (Lewisburg, PA: Bucknell University Press 1980), pp. 61–8.
33. Al-Bukhari, *al-Jami' al-Sahih*, p. 445; K: 67, B: 85.
34. R. Hall, S. James and J. Kertesz, *The Rapist Who Pays the Rent* (Bristol: Falling Wall Press, 1981).
35. Brownmiller, *Against our Will*, p. 17.
36. Ruth E. Hall, *Ask Any Woman*, (Bristol: Falling Wall Press, 1985), p. 32.
37. L.J.F. Smith, *Concerns About Rape*, Home Office Research Study No.106 (London: HMSO, 1989).
38. Bokhari, Vol. IV, p. 91, quoted in Bouhdiba, *Sexuality in Islam, p. 117.*
39. Delaney, *The Seed and the Soil*, p. 290.
40. Suyuti, quoted in Abdelwahab Bouhdiba, *Sexuality in Islam* (London: Routledge and Kegan Paul, 1985), p. 76.
41. Delaney, *The Seed and the Soil*, pp. 320–1.
42. Ibid., p. 296.

6   Women's liberation from incubator status – the story

1.   Cole, *Early Theories of Sexual Generation*, p. 200.
2.   Lyonet, quoted in Cole, ibid., p. 83.
3.   Senebier, quoted in Cole, ibid., p. 118.
4.   John Mason Goode, 'Generation' in *Pantologia, A New Cyclopaedia* (London: Kearsley Walker et al., 1813), Col. 5.
5.   Leonardo da Vinci, quoted in C.D. Darlington, *Genetics and Man*, (London: George Allen and Unwin, 1964), p. 41.
6.   P.L.M. Maupertuis, *Venus Physique* (Leyde, 1745); see also Roe, *Matter, Life, and Generation*, pp. 13–15.
7.   Baer, *Autobiography of Dr. Karl Ernst von Baer*, p. 218.
8.   Meyer, *Human Generation*, p. 73.
9.   Baer, *Autobiography of Dr. Karl Ernst von Baer*, p. 222.
10. Ibid., p. 217.
11. Ibid., p. 217.
12. Ibid., p. 225.

13. Ibid., p. 225.
14. Ibid., p. 226.
15. Ibid., p. 222.
16. C.D. Darlington, *Genetics and Man*, (London: George Allen and Unwin, 1964), p. 66.
17 Baer, quoted in Meyer, *Human Generation*, pp. 116–17.
18. Cole, *Early Theories of Sexual Generation*.

7   Woman the seed, man the stimulator

1. Colin M. Turnbull, 'Mbuti Womanhood', in F. Dahlberg (ed.), *Woman the Gatherer*, p. 209.
2. Colin M. Turnbull, *The Forest People* (New York: Simon and Schuster, 1961), p. 169.
3. Conway Zirkle, 'Animals Impregnated by the Wind', *Isis*, 1936, 25, pp. 95–6
4. Gimbutas, *The Language of the Goddess*, p. 218.
5. Gerda Lerner, *The Creation of Patriarchy*, p. 195.
6. E.O. James, *The Ancient Gods* (London: Weidenfeld and Nicolson, 1960).
7. Walker, *The Woman's Encylopedia of Myths and Secrets*, p. 976; see also E.O.G. Turville-Petre, *Myth and Religion of the North* (New York: Holt Rinehart and Winston, 1964), p. 120.
8. W. Robertson Smith, *Religion of the Semites* (London, A. and C. Black, 1894).
9. Malinowski, *The Sexual Life of Savages in North-Western Melanesia*, p. 162.
10. Ibid.
11. See Marilyn French, *Beyond Power: On Women, Men and Morals* (London: Abacus/Sphere, 1986), p. 58.
12. Barbara S. Lesco, *Biblical Archaeologist*, Vol. 54, No.1 (March 1991).
13. Cameron, *Daughters of Copper Woman*, p. 100.
14. Charles Seltman, *The Twelve Olympians* (London: Pan Books, 1952).
15. The passage from the *Oresteia* is quoted in full on p. 3. Aeschylus, *Plays Two. Oresteia: Agamemnon, Libation-Bearers, Eumenides*, p.114.
16. Ibid., p. 116.
17. Anthony J. Podlecki, *The Political Background of Aeschylean Tragedy* (Ann Arbor: University of Michigan Press, 1966) pp. 1, 7, 155.

8   An all-encompassing goddess or mere fertility cult?

1. Apuleius, *Golden Ass*, trans. Robert Graves (New York: Pocket Books, 1951).

2. Quoted in Evelyn Reed, *Woman's Evolution – from Matriarchal clan to patriarchal family* (Pathfinder, New York, 1975).
3. Allen, *The Sacred Hoop*, p. 15.
4. Elise J. Baumgartel, *The Cultures of Prehistoric Egypt* (London: Oxford University Press, on behalf of the Ashmolean Museum, Oxford, 1960), p. 71.
5. Colin Renfrew, 'Cycladic Art', *Down to Earth*, Factsheet No.1, (Channel 4 Television, 1992).
6. Baumgartel, *The Cultures of Prehistoric Egypt*, p. 71.
7. J.H. Luquet, *The Art and Religion of Fossil Man* (London, 1930), pp. 109–10.

9   The phallus

1. Gregersen, *Sexual Practices*, p. 293.
2. Davis, *The First Sex*, p. 99.
3. Ibid., p. 97–98.
4. Ibid., p. 97.
5. Gimbutas, *The Goddesses and Gods of Old Europe*, p. 237.
6. Gimbutas, *The Language of the Goddess*, p. 110.
7. A'nanda Mu'rti, *Discourses on Tantra*, Vol. 1.
8. Arthur Cotterell, 'The Indus Civilisation', in *The Encyclopedia of Ancient Civilizations*, p. 181.
9. Philip Rawson, *Oriental Erotic Art* (Ware, Herts: Omega, 1984), p. 29.
10. A'nanda Mu'rti *Discourses on Tantra*, Vol. 1, p. 154.
11. Ibid., p. 157.
12. Gimbutas, *The Language of the Goddess*, p. 231.
13. Ibid., p. 181.
14. Veen, *The Goddess of Malta* (Inanna-FIA Publications, 1992), p. 20.
15. J. Bezzina, *The Ggantija Temples* (Gozo, Malta: Xaghra), p. 4.
16. Ibid.

10   Woman the boss

1. Helen Diner, (pseud. Bertha Eckstein-Diner) *Mothers and Amazons: The First Feminine History of Culture*, (London, 1932), p. 170.
2. Ibid.
3. Quoted in Stone, *When God was a Woman*, p. 36.
4. Ibid., p. 35.
5. W. Boscawen, *Egypt and Chaldea* (London: Harper, 1894).
6. Barbara Lesco, *Biblical Archaeologist*, Vol. 54, No. 1 (March 1991) (Atlanta, GA: Scholars Press).
7. Herodotus, *The History*, p. 108 (Euterpe II:35).
8. Quoted in Stone, *When God was a Woman*, p. 45.
9. Davis, *The First Sex*, p. 177.

10. Charles Seltman, *The Twelve Olympians* (New York Apollo Editions, 1962), p. 27.
11. Lesco, *Biblical Archaeologist*.
12. Ibid.
13. Henri Frankfort, *Kingship and the Gods* (University of Chicago Press, 1948).
14. Walther Hinz, *The Lost World of Elam* (New York University Press, 1973).
15. James Mellaart, *Catal Huyuk, A Neolithic Town in Anatolia* (London and New York: Thames and Hudson, 1967).
16. H. J. Rose, *The Handbook of Greek Mythology* (London: Methuen, 1928).
17. Miles, *The Women's History of the World*, pp. 43–4.
18. Stylianos Alexiou, quoted in Stone, *When God was a Woman*, p. 144.
19. Stone, ibid., p. 135.
20. L. Frobenius, *The Childhood of Man* (London: Seeley, 1909).
21. Stone, *When God was a Woman*, pp. 142–3.
22. Ibid., p. 157.
23. Herodotus, *The History*, p. 972 (Melpomene IV:176).
24. Ibid., p. 272 (Melpomene IV:104).
25. Ehrenberg, *Women in Prehistory*, pp. 77–8.
26. Gimbutas, *The Language of the Goddess*, p. 321.
27. Sidney Smith, quoted in S.H. Hooke (ed.), *Myth, Ritual and Kingship* (Oxford University Press, 1958).
28. Stone, *When God was a Woman*, p. 82.
29. Miles, *The Women's History of the World*, p. 77.
30. Richard Mahler, *Matriarchy Research and Reclaim Newsletter*, No. 74 (Autumn Equinox 1991), reprinted from *Great Expeditions* (May–June 1991), p. 21.
31. Allen, *The Sacred Hoop*, p. 18.

11 Portrait of a takeover: Yahweh

1. Raphael Patai, *The Hebrew Goddess* (New York: Avon, 1978) pp. 12–13.
2. W.F. Albright, *Yahweh and the Gods of Canaan* (London: Athlone Press, 1968).
3. I. Epstein, *Judaism* (London: Penguin, 1959).
4. Ginzberg, *Legends of the Bible*, pp. 34–5.
5. Ibid., p. 35.
6. Riencourt, *Sex and Power in History*, pp. 37–8.
7. Lerner, *The Creation of Patriarchy*, p. 184–5.
8. Ginzberg, *Legends of the Bible*, p. 40.
9. Stone, *When God was a Woman*, p. 215.
10. Ginzberg, *Legends of the Bible*, p. 49.
11. Stone, *When God was a Woman*, pp. 212–13.
12. Ginzberg, *Legends of the Bible*, p. 616.

12   The fall

1. Sarpong, *Girls' Nubility Rites in Ashanti*, p. 8, (Note 16).
2. Ibid., p. 5.
3. Ibid., pp. 5–6.
4 Ibid., p. 6.
5. Ibid., p. 8.
6. Ibid., p. 7.
7. Walker, *The Woman's Encyclopedia of Myths and Secrets*, p. 502.
8. Schlegal, *Male Dominance and Female Autonomy*, pp. 141–2.
9. Ehrenberg, *Women in Prehistory*, p. 99.
10. Miles, *The Women's History of the World*, p. 46.
11. Ibid., p. 57.
12. Gimbutas, *The Language of the Goddess*, p. xx.
13. Riencourt, *Sex and Power in History*, p. 34.
14. Gimbutas, *The Language of the Goddess*, p. xx.
15. Ibid.
16. Fabre d'Olivet, 'Histoire Philosophique du Genre Humain', in Edouard Schure (ed.), *The Great Initiates*, Vol. 1, (New York: McKay, 1913), pp. 26–52.
17. H.W.F. Saggs, *The Greatness that was Babylon*, (New York: Mentor, 1968).
18. Aldred, *Egypt to the end of the Old Kingdom*, p. 45.
19. Nancy Tanner quoted in Schlegal, *Male Dominance and Female Autonomy*, p. 143.
20. Richard Mahler, *Matriarchy Research and Reclaim Network Newsletter*, No. 74 (Autumn Equinox 1991).
21. Ibid.
22. Richard Mahler, *Great Expeditions* (May–June 1991), p. 21.
23. Allen, *The Sacred Hoop*, p. 268.
24. Ibid., p. 32.
25. Ibid., p. 13.
26. Ibid., p. 41.
27. Ibid., pp. 27–8.
28. Ibid., p. 28.
29. Ibid., p. 26.
30. Ibid., p. 33.
31. Ibid.

13   The parthenogenetic woman

1. H. Vallois, 'The Social Life of Early Man: The Evidence of Skeletons', in S.L. Washburn (ed.), *The Social Life of Early Man*, (Chicago: Aldine, 1961), pp. 214–35.
2. W.C. McGrew, 'The Female Chimp as Evolutionary Prototype', in F. Dahlberg (ed.), *Woman The Gatherer*, p. 54, 57.
3. Abbé Henri Breuil, 'The Paleolithic Age', in Rene Huyghe (ed.), *Larousse Encyclopedia of Prehistoric and Ancient Art*, (London: Paul Hamlyn, 1962), p. 46.

4. Camilla Power, *The Woman with the Zebra's Penis – Evidence for the Mutability of Gender among African Hunter-gatherers,* MSc Dissertation, Dept of Anthropology, University College, London, 1993.
5. Allen, *The Sacred Hoop,* p. 28.
6. A.A. Yengoyan, 'Demographic and Ecological Influences on Aboriginal Australian Marriage Sections', in R. Lee and I. De Vore (eds), *Man the Hunter,* (Chicago: Aldine, 1968), pp. 185–99.
7. Rice, 'Prehistoric Venuses: Symbols of Motherhood or Womanhood?', *Journal of Anthropological Research* (1981), pp. 402.
8. G.H. Luquet, *The Art and Religion of Fossil Man* (London: 1930), pp. 109–10.
9. Gimbutas, *The Language of the Goddess,* p. xxii.
10. Ibid., p. 321.
11. Ehrenberg, *Women in Prehistory,* p. 73.
12. Ibid., p. 72.
13. Don Ringe, in 'Before Babel', *Horizon,* BBC TV, 6 April 1992 (transcript p. 23).
14. Ehrenberg, *Women in Prehistory,* pp. 73–4.
15. Ibid., p. 74.
16. Nick Douglas, *Sexual Secrets* (London: Hutchinson, 1979), p. 199.
17. Malinowski, *The Sexual Life of Savages,* p. 155.
18. Ibid., p. 156.

14   Woman the prime mover

1. Ehrenberg, *Women in Prehistory,* p. 50.
2. Charles Darwin, *The Descent of Man, and Selection in Relation to Sex* (1871), p. 901.
3. Irven de Vore, 'A Comparison of the Ecology and Behaviour of Monkeys and Apes', in S.L. Washburn (ed.), *Classification and human evolution,* (Chicago: Aldine, 1963), pp. 301–19; 'Male Dominance and Mating Behaviour in Baboons', in F. Beach (ed.), *Sex and Behavior,* (New York: Wiley, 1965), pp. 266–89.
4. Mary-Claire King and A.C. Wilson, 'Evolution at Two Levels in Humans and Chimpanzees', *Science,* 188 (1975), pp. 107–16; and Dorothy Miller, 'Evolution of Primate Chromosomes', *Science,* 198 (1977), pp. 1116–24.
5. Adrienne L. Zihlman, 'Women as Shapers of the Human Adaptation', in F. Dahlberg (ed.), *Woman the Gatherer,* p. 81.
6. Reed, *Sexism and Science,* p. 31.
7. Shirley C. Strum, 'Life with the Pumphouse Gang', *National Geographic* (May 1975), p. 680.
8. Ibid., see also W.C. McGrew, 'The Female Chimpanzee as Evolutionary Prototype, in F. Dahlberg (ed.), *Woman the Gatherer,* pp. 54, 57.

9. McGrew, 'The Female Chimpanzee as Evolutionary Prototype', ibid, p. 52.
10. Ibid.
11. Ibid.
12. Ibid., p. 43.
13. G. Gray Eaton, 'The Social Order of Japanese Macaques', *Scientific American* (October 1976), p. 102.
14. Phyllis Jay, 'The Indian Langur Monkey', in Charles H. Southwick (ed.), *Primate Social Behavior*, (Princeton, NJ: Van Nostrand, 1963), p. 121.
15. Jane Beckman Lancaster, 'Stimulus Response', *Psychology Today*, (September 1973), p. 34.
16. Zihlman, 'Woman as Shapers of the Human Adaptation', p. 81; Donald Stone Sade, 'A Longitudinal Study of Social Behavior of Rhesus Monkeys' in R.H. Tuttle (ed.), *The Functional and Evolutionary Biology of Primates* (Chicago: Aldine, 1972) pp. 378–98.
17. Jane Beckman Lancaster, 'Carrying and Sharing in Human Evolution'. *Human Nature*, Vol. 1, No. 2 (1978), pp. 82–9.
18. Caroline E.G.Tutin, 'Mating Patterns and Reproductive Strategies in a Community of Wild Chimpanzees (Pan troglodytes schwein-furthii)', *Behavioral Ecology and Sociobiology*, 6 (1979), pp. 29–38; 'Reproductive Behavior of Wild Chimpanzees in the Gombe National Park, Tanzania', *Journal of Reproductive Fertility*, Supplement 28 (1980).
19. Zihlman, 'Women as Shapers of the Human Adaptation', p. 102.
20. Ibid., p. 108.
21. Ibid., p. 109.
22. Lorna Marshall, *The !Kung of Nyae Nyae*, (Cambridge, MA: Harvard University Press, 1976).
23. D.J. Mulvaney, *The Prehistory of Australia*, 2nd edn (Victoria: Penguin, 1975); also Brian Hayden, 'Sticks and Stones and Ground Edge Axes: the Upper Paleolithic in Southeast Asia?' in J. Allen, J. Golson and R. Jones (eds), *Sunda and Sahul: prehistoric studies in Southeast Asia, Melanesia and Australia* (New York: Academic Press, 1977).
24. Zihlman, 'Women as Shapers of the Human Adaptation', p. 107.
25. Agnes Estioko-Griffin and P. Bion Griffin, 'Woman the Hunter: The Agta', in F. Dahlberg (ed.), *Woman the Gatherer*, pp. 143–4.
26. Ibid., p. 131.
27. Ibid., p. 140.
28. Colin Turnbull, 'Mbuti Womanhood' in F. Dahlberg (ed.), *Woman the Gatherer*, p. 212–14.
29. Ibid., p. 206.
30. Ibid., p. 207.
31. Ibid., p. 209.
32. Ibid., p. 208.
33. Ibid., p. 209.
34. Ibid., p. 211.
35. Ibid., pp. 214–15.

36. Ibid., p. 211.
37. Ibid., p. 218.
38. Ibid., p. 214
39. Ibid., p. 219.
40. Frances Dahlberg, 'Introduction', *Woman the Gatherer*, p. 26.
41. Catherine H. Berndt, 'Interpretations/Facts in Aboriginal Australia', in F. Dahlberg (ed.), *Woman the Gatherer*, p. 196.
42. James Woodburn, 'Social Organisation of the Hadza of North Tanganyika', PhD Thesis, University of Cambridge, 1964, pp. 298–9; Camilla Power, *The Woman with the Zebra's Penis – Evidence for the Mutability of Gender among African Hunter-gatherers*, MSc Dissertation, Dept of Anthropology, University College, London, 1993, pp. 16–20.
43. Dahlberg, 'Introduction', *Woman the Gatherer*, p. 27.

15   Trobriands – islands in the sun

1. Malinowski, *The Sexual Life of Savages in North-Western Melanesia*, p. 3, 2–3.
2. Ibid., p. 144.
3. Ibid., p. 3.
4. Ibid., p. 168.
5. Ibid., p. 146.
6. Ibid., p. 176.
7. Ibid., p. 152.
8. Ibid., p. 45–6.
9. Ibid., pp. 49, 48.
10. Ibid., p. 15.

Conclusion

1. House of Lords Bill 37, (London: HMSO, 1992).
2. *Women's Action*, Vol. 5 No. 1, (November 1993), p. 1 (Equality Now, P.O. Box 20646, Columbus Circle Station, New York 10023.) Also contact Forward UK, Africa Centre, 38 King Street, London WC2 8JT, tel: 071–379 6889.
3. Debbie Taylor/New Internationalist Collective, *Women, a World Report* (London: Methuen, 1985), p. 65.
4. UNICEF, *State of the World's Children Report 1994*.
5. John Stuart Mill, *The Subjugation of Women* (1869; reprinted London: Dent/Everyman's Library), p. 223.
6. Malinowski, *Sex and Repression in Savage Society*, p. 24.
7. *Horizon*, BBC TV, 11 April 1994.
8. Dr Peter Goodfellow and Dr Robin Lovell-Badge, *Science*, (9 May 1991).

Bibliography

Ackrill, J.R. (ed.), *A New Aristotle Reader* (Oxford University Press, 1987).

Aeschylus, *Plays Two. Oresteia: Agamemnon, Libation-Bearers, Eumenides*, trans. Frederic Raphael and Kenneth McLeish (London: Methuen, 1991).

Aldred, Cyril, *Egypt to the end of the Old Kingdom* (London: Thames and Hudson, 1988).

Allen, Garland, *Life Science in the Twentieth Century* (Cambridge University Press, 1978).

Allen, Paula Gunn, *The Sacred Hoop* (Boston: Beacon Press, 1986).

A'nanda Mu'rti, Shrii Shrii, *Discourses on Tantra*, Vol. 1, (Calcutta: A'nanda Ma'rga Publications, 1993).

Aristotle, *The Generation of Animals,* trans. A. Platt, Vol. 8 of *The Complete works of Aristotle*, ed. J. Barnes, (Princeton University Press, 1984).

Astruc, J., *A Treatise on the Diseases of Women*, Vol. III (London: J. Nourse Bookseller, 1762).

Austin, C.R. and R.V. Short, (eds), *Reproduction in Mammals*, Book 1: Germ Cells and Fertilisation, (Cambridge University Press, 1972).

—— *Reproduction in Mammals*, Book 8: Human Sexuality, (Cambridge University Press, 1980).

Baer, Karl Ernst von, *Autobiography of Dr. Karl Ernst von Baer*, ed. Jane M. Oppenheimer (Canton, MA: Watson Publishing International, 1986).

—— *De ovi mammalium et hominis genesi epistola* (Leipzig: L. Voss, 1827).

—— 'On the Genesis of the Ovum of Mammals and of Man', trans. Charles Donald O'Malley. Introduction by I. Bernard Cohen, *Isis*, 47 (1956): pp. 117–53.

Barlow, Nora (ed.), *The Autobiography of Charles Darwin* (London: Collins, 1958).

Battersby, Christine, *Gender and Genius* (London: Women's Press, 1989).

Bodemer, Charles W., 'Regeneration and the Decline of Preformationism in Eighteenth Century Embryology', *Bulletin of the History of Medicine* Vol. XXXVIII (Baltimore: The Johns Hopkins Press, 1964), pp. 20-31.

Bodemer, Charles W., 'Embryological Thought in Seventeenth-Century England', *Medical Investigations in Seventeenth Century England* (Los Angeles: University of California William Andrews Clark Memorial Library, 1968).

Bord, Janet and Colin, 'Fertility Practices in Pre-Industrial Britain', *Earth Rites* (London: Granada, 1982).

Bouhdiba, Abdelwahab, *Sexuality in Islam*, trans. Alan Sheridan (London: Routledge and Kegan Paul, 1985).

Bowler, Peter J., 'Bonnet and Buffron: Theories of Generation and the Problem of Species', *Journal of the History of Biology*, Vol. 6, No. 2 (Autumn 1973), pp. 259–81.

—— 'Preformation and Pre-existence in the Seventeenth Century: A Brief Analysis, *Journal of the History of Biology*, Vol. 4, No. 2. (Autumn 1971), pp. 221–44.

—— 'The Changing Meaning of Evolution', *Journal of the History of Ideas*, Vol. 36 (January–March 1975), pp. 95–114.

—— 'Darwin on Man in the Origin of Species: A Reply to Carl Bajema', *Journal of the History of Biology*, Vol. 22, No. 3. (Autumn 1989), pp. 497–500.

Boylan, Michael, 'The Galenic and Hippocratic Challenges to Aristotle's Conception Theory', *Journal of the History of Biology*, Vol. 17, No. 1 (Spring 1984), pp. 83–112.

Briffault, Robert, *The Mothers: A Study of the Origin and Sentiments and Institutions* (London: Allen and Unwin, 1952).

Brownmiller, Susan, *Against Our Will* (London: Penguin, 1976).

Cameron, Anne, *Daughters of Copper Woman* (London: Women's Press, 1984).

Campbell, Joseph, *The Masks of God: Occidental Mythology* (London: Arkana/Penguin, 1991).

—— *The Masks of God: Oriental Mythology* (London, Arkana/Penguin, 1991).

Carr, Donald E., *The Sexes* (London: Heinemann, 1970).

Chadwick, John, *The Decipherment of Linear B* (Cambridge University Press, 1958; reprinted 1990).

Cherfas, Jeremy and John Gribbin, *The Redundant Male: Is Sex Irrelevant in the Modern World* (London: Bodley Head, 1984).

Churchill, Frederick B., 'August Weisman and a Break from Tradition', *Journal of the History of Biology*, Vol. 1, No.1, (Spring 1968) pp. 91–112.

—— 'The History of Embryology as Intellectual History', *Journal of the History of Biology*, Vol. 3, No.1, (Spring 1970) pp. 155–81.

—— 'From Heredity Theory to Vererbung', *Isis*, 78 (1987), pp. 337–64.

—— 'Hertwig, Weismann and the Meaning of Reduction Division', *Isis*, Vol. 61 (1970), pp. 429–57.

—— 'Sex and the Single Organism : Biological Theories of Sexuality in the mid-Nineteenth Century', *Studies in the History of Biology*, Vol. 3, (1973), pp. 139–77.

Cole, Francis Joseph, *Early Theories of Sexual Generation* (Oxford: Clarendon Press, 1930).

Coleman, William, 'Cell, Nucleus and Inheritance: An Historical Study', *Proceedings of the American Philosophical Society*, Vol. 109 (1965), pp. 124–58.

Cooke, J., *The New Theory of Reproduction* (London: J. Buckland, 1762).

Cotterell, Arthur (ed.), *The Encyclopedia of Ancient Civilizations* (London: Macmillan, 1983).

Coutes, Irene, *The Seed Bearers – Role of the Female in Biology and Genetics* (Bishop Auckland: Pentland Press, 1993).

Croft, L.R., *The Life and Death of Charles Darwin*, (Chorley, Lancs: Elmwood, 1989).

Cronin, H., *The Ant and the Peacock: Altruism and Sexual Selection from Darwin to Today* (Cambridge University Press, 1991).

Dahlberg, Frances (ed.), *Woman the Gatherer* (New Haven: Yale University Press, 1981).

Davis, Elizabeth Gould, *The First Sex* (London: Penguin, 1979).

Delaney, Carol, *The Seed and the Soil* (Berkeley: University of California Press, 1991).

Dhonden, Dr Yeshi, 'Embryology in Tibetan Medicine', *Tibetan Medicine*, Series No. 1, (Library of Tibetan Works and Archives, 1980).

Dobell, Clifford, *Antony von Leeuwenhoek and his Little Animals* (New York: Dover, 1960).

Dowd, Maureen, 'When Men Get a Case of the Vapors', *New York Times*, 30 June 1991, p. 2E.

Ehrenberg, Margaret, *Women in Prehistory* (London: British Museum Publications, 1989).

Eisler, Riane, *The Chalice and the Blade* (San Francisco: Harper and Row, 1987).

El Saadawi, Nawal, *The Hidden Face of Eve* (London: Zed Books, 1980).

Engels, Frederick, *The Origin of the Family, Private Property, and the State* (New York: Pathfinder Press, 1972).

Farley, John, *Ideas about Sexual Reproduction 1750–1914: Gametes & Spores* (Baltimore and London: Johns Hopkins University Press, 1977).

Foote, Edward T. PhD, 'Harvey: Spontaneous Generation and the Egg', *Annals of Science*, Vol. 25 (London: Taylor and Francis, 1969), pp. 139–63.

Ford, Norman M., SDB, *When did I begin?* (Cambridge University Press, n.d).

Foucault, Michel, *The History of Sexuality*, Vol. 1, (London: Pelican, 1981).

French, Marilyn, *The War Against Women* (London: Penguin, 1992).

Fuller, R. Buckminster, *The Critical Path* (London: Hutchinson, 1983).

Galen, *On the Usefulness of the Parts of the Body*, trans. M. May (Ithaca: Cornell University Press, 1968).

Garden, George, 'A Discourse Concerning the Modern Theory of Generation', *Philosophical Transactions of the Royal Society*, Vol. XVII, (1691), pp. 474–83.

Gasking, Elizabeth R., *Investigations into Generation 1651–1828*, London, Hutchinson, 1967.

Geddes, Patrick and J.A. Thomson, *The Evolution of Sex*, (London: Walter Scott, 1889).

Gero, Joan M. and Margaret W. Conkey, *Engendering Archaeology* (Oxford: Basil Blackwell, 1991).

Gimbutas, Marija, *The Goddesses and Gods of Old Europe, Myths and Cult Images* (London: Thames and Hudson, 1982).

—— *The Language of the Goddess* (London: Thames and Hudson, 1969).

Ginzberg, Louis, *Legends of the Bible* (Philadelphia: Jewish Publication Society of America, 1956).

Glass, Bentley, 'Maupertuis, Pioneer of Genetics and Evolution', in B. Glass et al. (eds), *Forerunners of Darwin: 1745–1859* (Baltimore: John Hopkins Press, 1959).

Godelier, Maurice, *The Making of Great Men*, (Cambridge University Press, n.d.).

Goode, John Mason, *The Book of Nature* (1826).

Graham, Harvey, *Eternal Eve* (London: Heinemann Medical Books 1950).

Graves, Robert, *The White Goddess* (London: Faber and Faber, 1961).

Gregersen, Edgar, *Sexual Practices, The Story of Human Sexuality* (London: Mitchell Beazley, 1982).

Haeckel, Ernst, *The Riddle of the Universe (At the close of the Nineteenth Century)*, trans. Joseph McCabe, (Buffalo, NY: Prometheus Books, 1899).

Harding, Sandra, *Whose Science? Whose Knowledge?* (Milton Keynes: Open University Press, 1991).

Harris, Margaret, 'Official: Women doubly smart', *Sydney Morning Herald*, 5 June 1991, p. 3.

Harvey, William, *Disputations Touching the Generation of Animals*, trans. Gweneth Whitteridge (Oxford: Blackwell Scientific Publications, n.d.).

Herodotus, *The History*, trans. Henry Cary, (Buffalo: Prometheus Books, 1992).

Hertig, Arthur T. MD, 'A Fifteen-Year Search for First-Stage Human Ova', *Journal of the American Medical Association*, Vol. 261, No. 3 (20 January 1989), pp. 434–5.

Hertwig, Oscar, *The Biological Problem of Today: Preformation or Epigenesis? The Basis of a Theory of Organic Development* trans. P. Chalmers Mitchell, (London: Heinemann, 1896).

Hewson, M. Anthony, *Giles of Rome and the Medieval Theory of Conception* (London: Athlone Press, 1975).

Holmes, S.J., 'K.E. von Baer's Perplexities over Evolution', *Isis*, 37 (1947), pp. 7–14.

Horowitz, Maryanne Cline, 'Aristotle and Women', *Journal of the History of Biology*, Vol. 9, No. 2, (Autumn 1976), pp. 183–213.

Hubbard, Ruth, *The Politics of Women's Biology* (New Brunswick: Rutgers Press, 1990).

Jacob, Francois, *The Logic of Life: A History of Heredity*, trans. Betty E. Spillmann, (London: Allen Lane, 1973).

Jaykar, Pupul, *The Earth Mother* (London: Penguin, 1989).

Johns, Catherine, *Sex or Symbol, Erotic Images of Greece and Rome* (London: British Museum Publications, 1982).

Johnson, Martin, and Barry Everitt, *Essential Reproduction*, 3rd edn (Oxford: Blackwell Scientific Publications, 1988, 1991).

Kennon, Donald R., 'An Apple of Discord: The Woman Question at the World's Anti-Slavery Convention of 1840' *Slavery & Abolition*, Vol. 5, No. 3 (December 1984).

Knight, C.D., *Blood Relations. Menstruation and the Origins of Culture* (New Haven and London: Yale University Press, 1991).

Laqueur, Thomas, *Making Sex: Body and Gender from the Greeks to Freud* (Cambridge MA: Harvard University Press, 1990).

Leeuwenhoek, Antoni van, *Philosophical Transactions of the Royal Society*, Vol. XXV (1685), pp. 1120–34.

Lefkowitz, Mary R. and Maureen B. Fant, *Women's Life in Greece and Rome* (London: Duckworth, 1982).

Lerner, Gerda, *The Creation of Patriarchy*, (Oxford University Press, 1986).

Long, Asphodel P., *In a Chariot Drawn by Lions: The Search for the Female in Deity* (London: Women's Press, 1992).

Longo, Frank J., *Fertilisation* (London: Chapman and Hall, 1987).

Lowndes Sevely, Josephine, *Eve's Secrets* (London: Bloomsbury Publishing, 1987).

McConahay, Shirley A., and John B. McConahay, 'Sexual Permissiveness, Sex-Role Rigidity, and Violence across Cultures', *Journal of Social Issues*, Vol. 33, No. 2 (1977).

McLaren, Angus, *A History of Contraception* (Oxford: Basil Blackwell, 1990).

—— *Reproductive Rituals: The Perception of Fertility in England from the Sixteenth Century to the Nineteenth Century* (London: Methuen, 1984).

Malinowski, Bronislaw, *The Sexual Life of Savages in North-Western Melanesia*, 3rd edn (London: George Routledge and Sons, 1939).

—— *Sex and Repression in Savage Society*, (1927; reprinted London: Routledge and Kegan Paul, 1979).

Mernissi, Fatima, *Beyond the Veil* (London: Al Saqi, 1985).

—— *Women in Moslem Paradise* (New Delhi: Kali for Women, 1986).

Meyer, William Arthur, *The Rise of Embryology* (London: Oxford University Press, 1939).

—— *Human Generation: Conclusions of Burdach, Dollinger and von Baer* (Stanford University Press, 1956).

—— 'A Summary of von Baer's Commentary', *Bulletin of the Institute of the History of Medicine*, Vol. 1 (1938), pp. 1031–40.

Miles, Rosalind, *The Women's History of the World* (London: Paladin/Grafton, 1989).

Minces, Juliette, *The House of Obedience* (London: Zed Books, 1982).

Moore, K.L., *Essentials of Human Embryology* (Oxford: Blackwell Scientific Publications, 1988).

Morgan, Elaine, *The Descent of Woman* (London: Souvenir Press, 1985).

Morgan, Fidelis, *A Misogynist's Source Book* (London: Jonathan Cape, 1989).

Morsink, Johannes, 'Was Aristotle's Biology Sexist?' *Journal of the History of Biology*, Vol. 12, No. 1 (Spring 1979), pp. 82–112.

Needham, Joseph, *A History of Embryology* (Cambridge University Press, 1959).

Neumann, Erich, *The Great Mother, an Analysis of the Archetype*, trans. Ralph Manheim (London: Routledge and Kegan Paul, 1963).

Nilsson, Lennart, *A Child is Born* (London: Doubleday, 1990).

O'Faolain, Julia and Lauro Martines, *Not in God's Image* (London: Temple Smith, 1973).

Oppenheimer, Jane, 'K.E. von Baer's Beginning Insights into Causal-Analytical Relationships during Development', *Developmental Biology*, Vol. 7 (1963), pp. 11–21.

—— *Essays in the History of Embryology and Biology* (Cambridge, MA: MIT Press, 1967).

Oppong, Christine, *Marriage among a Matrilineal Elite* (Cambridge University Press, n.d.).

Ospovat, Don, 'The Influence of Karl Ernst von Baer's Embryology, 1828–1859: A Reappraisal in light of Richard Owen's and William B. Carpenter's '"Palaeontological Applications of von Baer's Law"', *Journal of the History of Biology*, Vol. 9, No.1 (Spring 1976), pp. 1–28.

Pagels, Elaine, *The Gnostic Gospels* (London: Pelican, 1982).

Payne, Alma Smith, *The Cleere Observer: A Biography of Antoni van Leeuwenhoek* (London: Macmillan, 1970).

Peter, Prince, HRH of Greece and Denmark, *A Study of Polyandry* (The Hague: Mouton and Co.,1963).

Pirani, Alix, *The Absent Mother, Restoring the Goddess to Judaism and Christianity* (London: Mandala/HarperCollins, 1991).

Preus, Anthony, 'Science and Philosophy in Aristotle's Generation of Animals', *Journal of the History of Biology*, Vol. 3, No. 1, (Spring 1970), pp.1–52.

—— 'Galen's Criticism of Aristotle's Conception Theory', *Journal of the History of Biology*, Vol. 10, No. 1 (Spring 1977), pp. 65–85.

Price, Dorothy, *A Historical Review of Embryology and Intersexuality Fact and Fancy* (Leiden: E.J. Brill, 1967).

Punnett, R.C., 'Ovists and Animalculists', *American Naturalist*, Vol. 62, pp. 481-507.

Quale, Robina, G., *A History of Marriage Systems* (New York: Greenwood Press, 1988).

Reed, Evelyn, *Problems of Women's Liberation* (New York: Pathfinder, 1970).

—— *Sexism and Science*, 3rd edn (New York, Pathfinder, 1989).

Reich, Wilhelm, *The Invasion of Compulsory Sex-Morality* (London: Souvenir Press, 1972).

Renfrew, Colin, *Archaeology and Language* (London: Penguin,1989).

Rice, Patricia C., 'Prehistoric Venuses: Symbols of Motherhood or Womanhood?', *Journal of Anthropological Research* (1981), pp. 402–14.

Ridley, Matt, *The Red Queen: Sex and the Evolution of Human Nature* (London: Viking/Penguin, 1993).

Riencourt de, Amaury, *Sex and Power in History* (New York: Delta/Dell Publishing, 1974).

Roe, Shirley A., 'Rationalism and Embryology: Caspar Friedrich Wolff's Theory of Epigenesis', *Journal of the History of Biology*, Vol. 12, No. 1 (Spring 1979), pp. 1–43.

—— *Matter, Life, and Generation: Eighteenth-Century Embryology and the Haller–Wolff Debate* (Cambridge University Press, 1981).

Rome, Lucienne and Jesus, *Primitive Eroticism*, trans. Evelyn Rossiter (Ware: Omega Books, 1983).

Ruestow, Edward G., 'Images and Ideas: Leeuwenhoek's Perception of the Spermatozoa', *Journal of the History of Biology,* Vol. 16, No.2 (Summer 1983), pp. 185–224.

Russett, Cynthia Eagle, *Sexual Science: The Victorian Construction of Womanhood* (Cambridge, MA: Harvard University Press, 1989).

Saffron, Lisa, *Challenging Conceptions* (London: Cassell, 1994).

Sarpong, Rt Rev. Dr Peter, *Girls' Nubility Rites in Ashanti* (Tema: Ghana Publishing Corporation, 1977).

Sarton, George, 'The Discovery of the Mammalian Egg and the Foundation of Modern Embryology', *Isis,* 16, (1931) pp. 315–30.

Schlegel, Alice, *Male Dominance and Female Autonomy, Domestic Authority in Matrilineal Societies* (Hraf Press, 1972).

Schullian, Dorothy M., 'Notes and Events; Two Letters on Epigenesis from John Turberville Needham to Albrecht von Haller', *Journal of the History of Medicine* (January 1976), pp. 68–77.

Shorter, Edward, *A History of Women's Bodies,* (London: Pelican, 1984).

Shuttle, Penelope and Peter Redgrove, *The Wise Wound, Menstruation and Everywoman,* (London: Penguin, 1980).

Smith, John Maynard, *Did Darwin get it Right?* (London: Penguin, 1993).

Stone, Merlin, *When God was a Woman* (San Diego: Harcourt Brace Jovanovich, 1976).

Suzuki, David and Peter Knudtson, *Genethics: The Ethics of Engineering Life* (Toronto: Stoddart, 1988).

Tannahill, Reay, *Sex in History* (London: Sphere Books, 1981).

Thomson, Sir J. Arthur, and Patrick Geddes, *Life; Outlines of General Biology,* Vol. 1 (New York: Harper and Brothers, 1931).

Tuana, Nancy (ed.), *Feminism and Science* (Bloomington: Indiana University Press, 1989).

Veen, Veronica, *The Goddess of Malta: The Lady of the Waters and the Earth* (Haarlem: Inanna-Fia, 1992). (Available from Inanna-Fia Publications, P.O. Box 2017, 2002 CA Haarlem, Holland.).

Vetterling, Mary, Frederick A. Elliston, and Jane English, (eds), *Feminism and Philosophy* (Totowa: Littlefield Adams, 1977).

Walker, Barbara G., *The Woman's Encyclopedia of Myths and Secrets* (San Francisco, Harper and Row,1983).

Webb, Peter, *The Erotic Arts* (London: Secker and Warburg, 1983).

Weigle, Marta, *Creation and Procreation* (Philadelphia: University of Pennsylvania Press, 1989).

Wembah-Rashid, J.A.R., *The Ethno-History of the Matrilineal Peoples of Southeast Tanzania,* Acta Ethnologica et Linguistica No. 32 (Vienna, 1975).

Westermarck, Edward, *The History of Human Marriage,* Vol. 1 (London: Macmillan, 1921).

Whitmont, Edward C., *Return of the Goddess* (London: Arkana, 1982).

Woolley, Sir Leonard, *UR 'of the Chaldees'* (London: Herbert Press, 1929; revised edn with P.R.S. Moorey, 1982).

Zirkle, Conway, 'Animals impregnated by the Wind', *Isis,* 25, (1936), pp. 95–130.

Index

217